AF324189

Field Methods for Petroleum Geologists

Fakhry A. Assaad

Field Methods for Petroleum Geologists

A Guide to Computerized
Lithostratigraphic Correlation Charts
Case Study: Northern Africa

 Springer

Fakhry A. Assaad
1350, 37th, Str. East Tuscaloosa,
AL 35405
USA
FekAsa@aol.com

ISBN: 978-3-540-78836-2   e-ISBN: 978-3-540-78837-9

Library of Congress Control Number: 2008934328

Cover design: deblik, Berlin

Printed on acid-free paper

9 8 7 6 5 4 3 2 1

springer.com

Preface

The author worked at Sonatrach of Algeria (Societé National de Transport et de Hydrocarbures) for seven consecutive contracts over more than a decade, being in charge of the Triassic Province of Algeria, besides having responsibility for the Database Retrieval System (PetroData System), established in 1970 by the Exploration Directorate.

The author started his career with Sonatrach as a District Geologist Engineer in northern Algeria, and later worked with an international group from Technoexport (1971) on the Triassic Province of the Algerian Saharan platform.

Several projects were accomplished:

1. In 1970–1972, he carried out detailed structural and stratigraphic studies of the Triassic reservoirs, and defined the limits of deposition of the Triassic basin in Algeria.
2. Later, he established a computer application for a lithostratigraphic correlation chart of the sedimentary formation $(3\frac{1}{2} \times \frac{1}{2})$ m, as a type model for the Arabian Maghreb region (northwestern Africa), which can be further extended to cover North Africa.
3. He was in charge of the Database Retrieval System (PetroData System), as a liaison with the exploration districts to supervise and manage the quality of the input well data, as well as the geological, drilling, productive, and geophysical data, etc., in order to define and formulate retrieval requests for computer extraction procedures. He provided corrections and improvements to refine the edit, validation, and cross check functions of the maintenance subsystem.

Acknowledgement

The author is quite grateful for the quick response from Dr. Khalil Chekib, the Algerian Minister of Petroleum and Energy, and Dr. H. Chebourou, the Director of "Etudes et Synthése" of the Exploration Division of Sonatrach, Algiers, for updating documents necessary to the book.

My highest gratitude to my college's geologist Dr. Ali Werwer, the chairman and CEO of the Research Groundwater Co., (REGWA), Cairo; and Professor Dr. Hamed Metwalli, of the Petroleum Geology Department, Faculty of Science, Cairo University, for their sincere help in providing important materials for the book.

My special gratitude to Mr. John Sandy, the director of the Rodgers Library for Science and Engineering, University of Alabama; and to Mr. Lewis S. Dean, the head of the Library of the Geology Survey of Alabama (GSA), for providing me promptly with the needed references.

Contents

Part II   A Guide to Lithiostratigraphic Correlation Charts – A Model Type/North Africa

List of Figures

List of Tables

Part I
Petroleum Exploration Methods

Chapter 1

Introduction—Petroleum Hydrocarbons

1.1 Historical Aspect

The early uses of oil date from the time of Noah, who had used asphalt to make his ark watertight; in those times, oil was used for medication, waterproofing, and in warfare. Many references are found to using pitch or asphalt, collected from the natural seepage with which the Middle East abounds.

It is interesting to note that in early petroleum drilling operations, a major American company once employed a chief geologist whose exploration philosophy was to drill in old Indian graves; whereas another oil finder in the early 1900s used to put on an old hat, gallop about the prairie until his hat dropped off, and start drilling where it landed. History records that he was very successful!

1.2 Petroleum Occurrence and Chemical Composition

Petroleum is a naturally occurring complex mixture of hydrocarbons which may be either solid, liquid, or gas, depending upon its own unique composition and the pressure and temperature at which it is confined. The liquid phase of crude petroleum exists naturally in underground reservoirs and remains liquid at atmospheric pressure after passing through surface separation facilities. Petroleum may contain numerous impurities such as carbon dioxide, hydrogen sulfide, and other complex compounds of nitrogen, sulfur, and oxygen. Because of its numerous impurities it difficult to perform a precise chemical analysis; petroleum is therefore often classified by its base designation, e.g., paraffin base, asphalt base, or mixed base crude. Paraffin base crude is oil consisting mainly of paraffins, which leave a solid residue of wax when completely distilled.

Asphalt base crude is oil primarily composed of cyclic compounds (mostly naphthenes), which leave a solid residue of asphalt when distilled. Mixed base oils are those that fall in the middle of the other categories (Gatlin 1960).

There are three principal hydrocarbon series of petroleum:

(a) **Paraffins, or saturated hydrocarbons (alkanes)**, compounds of the general formula C_nH_{2n+2}, are chemically stable and have either straight or branched chains. All crude oils contain some paraffins, particularly as the more volatile (low boiling point) constituents.

   Fig. 1.1a is an example of the first few members of the paraffin series.

(b) **Cycloparaffins, or naphthenes**, of the general formula C_nH_{2n}, have a ring structure, e.g., cyclopropane and cyclobutane (Fig. 1.1b).

(c) **Aromatics or benzene series**, of the formula C_nH_{2n-6}, are chemically active and contain the benzene ring, the simplest member being benzene (Fig. 1.1c).

Many crude oils contain porphyrins, and nearly all contain nitrogen. Petroleum rotates the plane of polarized light; this property is primarily restricted to organic materials known as optical isomers; it further suggests the organic origin of petroleum, which accounts for the large quantities of carbon and hydrogen needed to form petroleum deposits. The role of anaerobic bacteria may play a part in promoting the

F.A. Assaad, *Field Methods for Petroleum Geologists*,
© Springer-Verlag Berlin Heidelberg 2009

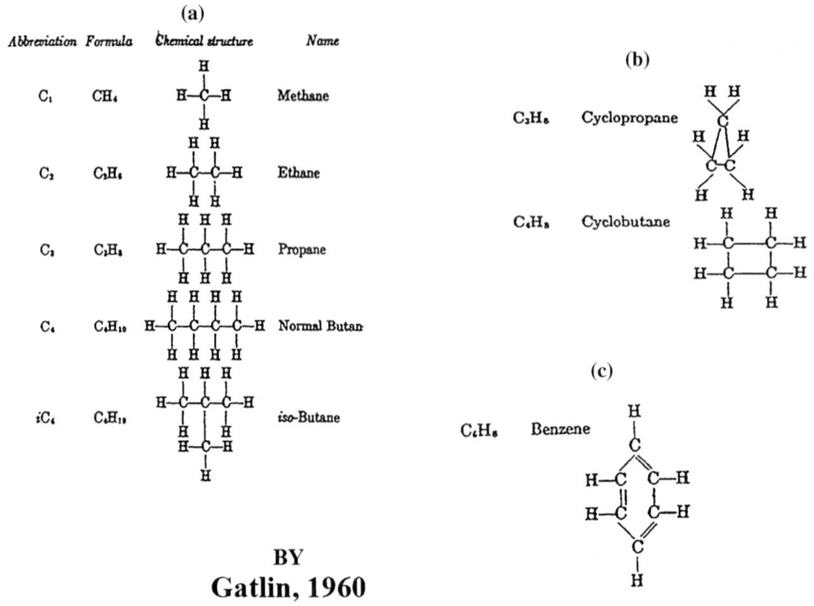

Fig. 1.1 (a) First few members of Paraffins Series. (b) Cycloparaffins. (c) Benzene Ring

BY

Gatlin, 1960

alteration process through which organic materials are transformed into petroleum.

1.3 Properties of Crude Oils

The crude oils range widely in their physical properties, e.g., color, specific gravity (0.80–0.95), and in their chemical properties. Gravity is the most effective energy causing the fluid's movement, although earth compression and capillarity also play a part. The difference in specific gravities separates the gas, oil, and water trapped below the seal or cap rock. According to the American Petroleum Institute (API), gravity is actually a measure of oil's density, and is related to specific gravity by the formula:

API gravity (in degrees) $= 141.5 - (131.5)$

(NB: API of 10° equivalent to sp. gr. of one). Sp. gr.

1.4 Natural Gas—Definition

Petroleum gas, or natural gas, recently became a highly valuable product. Previously, gas produced with oil was sold primarily on a local scale, and the excess was flared. As the natural gasoline and liquefied petroleum gas (butane and propane) industry developed, the utilization of the residue gas (dry gas remaining after liquids are removed) also increased. Natural gas and its associated products are virtually as important as oil.

The content of the natural gasoline or the liquefied petroleum gas can be expressed in GPM (gallons per thousand [Latin *mille*] standard cubic feet [MCF]); if gasoline's GPM = 1–2, it is considered wet; and if it is = 0.2, it is considered somewhat dry. Gas gravity, the ratio of the density of a gas to the density of air at standard conditions, is also used as a measurement.

Natural gas is commonly measured in cubic feet; for example, if we start with a given volume of gas at a temperature of 60°F and a pressure of one atmosphere (14.7 PSI at sea level), then an increase in temperature will cause the volume of the gas to increase. Gas volumes are written in multiples of 1,000, abbreviated as M; thus, 3,555,000 cubic feet of gas would be equal to 3,555 MCF.

It is worth mentioning that the migration of gas globules would be affected by the same factors as with oil globules. Since hydrocarbon gas is soluble in water, the solubility rises with an increase in pressure. An increase in the depth of burial will drive more gas into solution in any associated oil, and vice versa.

1.5 Petroleum Hydrocarbon Non-Reservoir Rocks

The hydrocarbons found in the non-reservoir sediments presumably are produced directly, either wholly or partly, from the hydrocarbons (organic matter) that are found in living plant and animal matter. The reservoir rock is unlikely to be the source rock. In such cases, there might be two stages for the development of petroleum hydrocarbons: (a) movement from the source rock to the reservoir rock (primary migration), and (b) movement and segregation within the reservoir rock (secondary migration).

The petroleum hydrocarbons deposited in non-reservoir shales and carbonates as disseminated soluble hydrocarbon particles, possibly of colloidal or microscopic size, were thus associated with the nonsoluble organic matter. By the time diagenesis was complete, most of the petroleum hydrocarbons were probably in the form of petroleum. During and after diagenesis, water was squeezed out of the non-reservoir rocks into the reservoir rocks, and a fraction of the petroleum and petroleum hydrocarbons was entrained in the water. The movement from the non-reservoir rocks to the reservoir rocks is called the primary migration to distinguish it from the concentration and accumulation into pools of oil and gas called the secondary migration.

Carbonates and shale with an originally high content of bituminous matter become good source rocks if they do not retain their hydrocarbon potentials and are subjected to alteration into a different type of rock under high temperature and pressure. However, hydrothermal solutions of high temperatures allow the carbonate rocks to be crystallized from impervious bituminous into a porous reservoir trap rock. The consequent normal epigenetic changes of clay and shale and the alteration of carbonate rocks provide the time to develop the mother rock of hydrocarbons that migrates to, or is retained in the nearby reservoir trapped rocks (Levorsen 1954).

1.6 Petroleum Reservoir Rocks

Petroleum reservoir is the portion of the rock that contains the pool of petroleum and consists of four essential elements: (1) the reservoir rock (pay sand, oil sand, or gas sand); (2) the pore space, or porosity, or the "effective pore space"—that portion of the reservoir rock available for the migration, accumulation, and storage of petroleum; (3) the fluid content (water, oil, and gas) occupied by the effective pore space, which may be in a state of either static or dynamic equilibrium in which petroleum then occurs within a water environment; the distribution of fluids in a reservoir rock being dependent on the densities of the fluids and the capillary properties of the rock; (4) the reservoir trap that holds oil and gas in place in the pool. Rock traps are formed from a wide variety of combinations of structural and stratigraphic features of the reservoir rocks. The trap generally consists of an impervious cover, overlying and sealing reservoir rock, e.g., impervious shales, salt, or anhydrite deposits.

A petroleum reservoir rock must have two completely different properties: (1) porosity (the fluid-holding capacity) of variable types of openings, e.g., the intergranular pores in the sedimentary rocks, cavities in fossil rocks such as the oolitic limestone, fractured rocks, and joints formed by solution; and (2) permeability (fluid-transmitting capacity).

1.7 Petroleum Migration and Accumulation

1.7.1 Introduction

The process of concentration and emplacement of the hydrocarbon accumulation is referred to as migration and accumulation. Petroleum and petroleum hydrocarbon migrate from the source rock into porous and permeable beds where they accumulate and continue their migration until finally trapped.

There are several forces that cause such migration: (a) **compaction** of sediments as the depth of burial increases; (b) **diastrophism** of crustal movements that cause pressure differentials and consequent subsurface fluid movements; (c) **capillary forces** that cause oil to be expelled from fine pores of source rocks by the preferential entry of water; and (d) **gravity forces** that promote fluid segregation because of the density factor.

Forces due to compaction and capillarity influence fluid movement in horizontal strata into the porous rocks, but differences of pressure occur when the strata become tilted and cause fluid movement from localities with higher pressure to those with lower pressure.

difference in fluid potential and the transmissibility of the aquifer.

1.7.2 Oil Migration

The **primary migration** of hydrocarbon takes place "in solution" in the pore space or adsorbed on organic or inorganic matter during its path from the source rock to the reservoir rock; in other words, petroleum hydrocarbons are entrained in the water that is squeezed out of the shales and clays during diagenesis and, to a lesser extent, in the confined water of the normal hydraulic circulation after diagenesis (Levorsen 1954). The **secondary migration** occurs when both movement and segregation of hydrocarbons occur within the reservoir rock. The effect of the hydrodynamic conditions can be extremely important to the movement of the petroleum.

Dilution by the entry of meteoric waters is improbable as an explanation for most oil accumulations; the migrating solution, on entering the reservoir rock, intermingles with water already in that rock, and a soap solution is effectively diluted and leads to the release of oil.

Because nearly all petroleum pools in oil producing regions exist within an environment of the confined (formation) water, petroleum migration is directly related to hydrology, fluid pressure, and water movement. The interstitial water content of the reservoir rocks generally shows a hydrodynamic fluid pressure gradient and moves in the direction of the lower fluid potential; the rate varies with the magnitude of the

1.7.3 The Role of Connate Water

At the time of complete diagenesis, reservoir and non-reservoir rocks are filled with confined (connate) water in which regional circulation patterns develop. This might be static if no fluid potential exists, but might continue to change if fluid pressure gradients change due to diastrophism, mountain building, erosion, deposition, etc. The flow of the connate or confined water may be either up or down the dip or slope, because the rate and direction of flow of confined water is proportional to the rate of change of the head of the hydrodynamic fluid potential gradient (measured in relation to sea level). This is not proportional to the rate of change in hydrostatic pressure along the flow path, because it can flow from an area of low fluid pressure to one of high fluid pressure, provided the head is lower in the direction of the flow. Fig. 1.2 is a diagrammatic profile along the direction of flow and pressure gradient of confined water (Levorsen 1954).

1.7.4 Differential Entrapment of Petroleum Hydrocarbons— Gussow Theory

Under ideal lithological and structural circumstances with widespread permeable rocks, Gussow's theory of

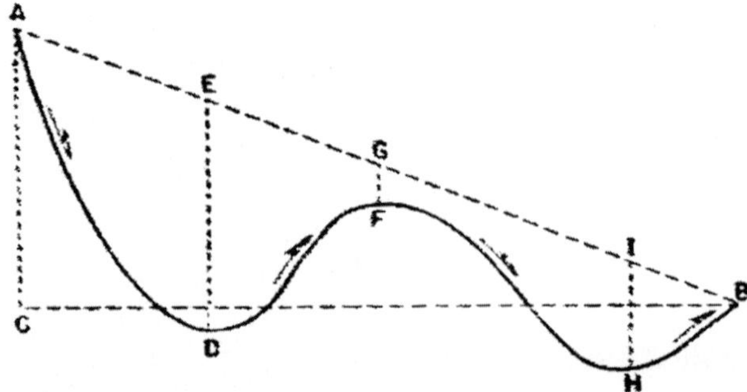

Fig. 1.2 Diagrammatic sketch showing the path water takes in flowing from intake area A to outlet area B as it passes across synclines and anticlines. The reservoir pressure at F would raise the water to G; this fluid pressure is less than at D, where the fluid pressure would raise the water to E, or at H, where the fluid pressure would raise the water to I. The surface AB is the potentiometric surface; its slope governs the overall flow of water from A to B. (Levorsen, 1954)

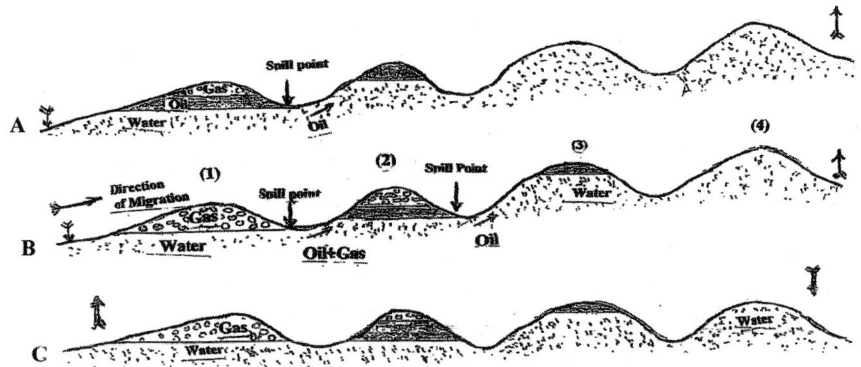

Fig. 1.3 Gussow Theory depends on differential entrapment of petroleum. In A & B, oil and gas moves from left to right up the regional dip, where oil is forced out at the spill point and is caught at the next trap; in trap (1): full of gas; trap (2): both of oil and gas; trap (3): contains only oil and the next trap (4): full of connate water (adapted from Levorsen, 1954)

differential entrapment explains how oil and gas were separated during updip regional migration through a series of structural traps. Gussow's theory discusses the occurrence of oil and gas in a nearby fold trap as oil is eventually forced out at the bottom of the fold, either because additional oil and gas enter the trap or because a loss of pressure brings about an increase in gas volume. When the trap is full to the spill point, an additional volume of oil or gas will force the excess oil out at the bottom of the fold, and will move up the dip and be caught in the next trap. This procedure is repeated, and oil continues to migrate and be trapped, until the final up-dip trap contains connate water. The theory of differential entrapment is a useful stimulus to explanation thinking, particularly when information from bore holes becomes available (Fig. 1.3, adapted from Levorsen 1954).

A **trap** is an area of low potential energy toward which the buoyant oil and gas move from areas of high energy, either aided or hindered by the movement of water in the aquifer. The oil and gas separate according to their differential density and remain in the trap, because the geologic environment (such as faulting, folding, or the rate of water flow through the reservoir rocks) does not change.

1.7.5 Petroleum Accumulation

There are many factors that lead to secondary migration of petroleum hydrocarbons through the reservoir rocks and their accumulation into pools, such as the

entrained particles, their buoyancy, tilted oil/water contacts, stratigraphic barriers, vertical migration, etc.

There is a wide variation in the character of the trap that may occur because of structural or stratigraphic factors, or a combination of both.

Structurally, entrapped petroleum can be released in either of two different ways or a combination of both: (a) the faulting up of reservoirs against impervious sections; and/or (b) the accumulation of hydrocarbons into local or regional apexes of the newly created reservoir's updip against unaltered impervious rocks.

Should the reservoir rock become tilted, there will be a greater chance of another migration (a tertiary migration), because the buoyancy factor due to the overall height of the oil or gas masses will be increased. Secondary migration of oil and gas can be provided by reducing the hydrostatic head through erosion. Tertiary migration, and ultimately oil loss, might occur during the active structural history of an area.

A **stratigraphic trap** in a reservoir rock may occur in a structural oil field or a gas field, sealed by an impermeable barrier, which has been tilted by an earth movement, but not folded or faulted in such a way that could affect the accumulation process. An isolated offshore sand bar completely surrounded by muddy sediments may contain a stratigraphic accumulation if hydrocarbons have entered it from these muddy sediments and are unable to migrate further. A coral reef can form a stratigraphic trap if it is surrounded by muddy sediments and sealed by an impermeable muddy cap rock.

Stratigraphic traps may be divided into two types: primary stratigraphic traps, which form during the deposition or diagenesis of the rock, including those formed by lenses, facies changes, and reefs; and secondary stratigraphic traps, which result from later causes such as solution and cementation, but chiefly from unconformities.

The lower boundary of the reservoir, either wholly or partly, is the plane of contact (oil–water or gas–water contact), simply given as O/W or G/W.

Leakages of hydrocarbons from accumulations are referred to as seepages or shows and occur either by exposure of the reservoir rock as a result of erosion, or from a buried reservoir rock that faces "permeable" faults, or joints which allow fluid movement through the cap rock. In such cases, although the reservoir rock is exposed, asphaltic matter from the crude oil in the reservoir rock might clog the rock pores at the surface, thereby greatly reducing the rate of loss of oil (Levorsen 1954). On the other hand, gas might escape under water; and in some cases both gas and water move through clayey rocks, forming a gas-charged mud, that issues as a mud volcano on the surface.

1.7.5.1 Other Factors Affecting Oil Accumulation

Time Factor: The time needed to form an accumulation depends on many factors, including the rate at which oil is entering the reservoir rock, and the time at which the release of oil occurs. Lateral and upward movements result in a better local accumulation., It took approximately one million years for oil and gas to originate and migrate, but it might take a much shorter time to accumulate into pools, possibly only thousands or even hundreds of years.

Temperature and Pressure Factors: The temperature of a reservoir may fluctuate and can be approximately 75°C (167°F), with a maximum of 163°C (325°F) in the sediments of an oil region. The **fluid pressures** within the reservoir rock may also fluctuate during the life of the reservoir, depending on the geologic history of the region, and might range from one atmosphere up to 1,000 atmospheres. The geologic history of the trap may vary widely—from a single geologic episode to a combination of many phenomena extending over a long period of geologic time.

An **increase in the depth of burial** will drive more gas into solution in any associated oil, and vice versa. Natural gas is much more soluble in oil than in water. However, a reduction in the depth of burial may, by releasing gas from solution in the oil, cause the vol-

ume of the hydrocarbons to exceed the capacity of a trap. Excess hydrocarbons would move from the trap to escape or to accumulate elsewhere; the possibility of hydrocarbon-release by increase in salinity may also occur in many oilfields; the brines are more saline than the normal seawater; if such a feature develops early in the history of the rocks, then as regards time, it would be compatible with the requirement that the oil should appear at an early date in the reservoir rock.

Fluids during compaction move upwards through the sediments, where their exact path is determined by the geometry and permeability of the encountered beds. When an oil mass or gas mass of suitable height has been built up, it will be capable of rising because of its buoyancy within the reservoir rock; and the movement could be aided by favorably directed water flow.

The natural energy that is present in the reservoir and available to move the oil into the producing wells is the potential energy of reservoir pressure, which is stored mainly in the compressed fluids, and its amount depends largely on the fluid potential, or head, of the reservoir fluids; and to a lesser extent on the compressed rocks that form the reservoir rock (Levorsen 1954).

1.8 The Capacity of Oil and Gas Traps

The law of gases discusses how the volume of a gas varies in direct proportion to the absolute temperature, but inversely with pressure ($V \propto T/P$); natural gases can be trapped in a pool in a separate reservoir and under a single pressure system. The capacity of a trap to hold gas is a direct function of the reservoir pressure; i.e., when a trap is completely full of gas, the present accumulation of gas could not have occurred until the present pressure was reached. Because the reservoir pressures in most areas are related to the depth of burial, the trap could not have been filled until the present depth of overburden was deposited. In a trap full of gas at a depth of 5,000 ft (1,592.4 m), the gas could not all have entered the trap until its pressure had reached 143 atmospheres (the average pressure at that depth); in a lower pressure environment, less gas would have been required to fill the trap.

Oil has its greatest mobility when its viscosity is the lowest and its buoyancy is the highest—in other words, at its saturation pressure. Upward movement might take place through well-developed structures of reservoir rocks because of the buoyancy of the fluid or because of the density factor. More gas was formed in place from oil or organic matter, being evolved at the expense of the heavier hydrocarbons by biochemical, thermal, or catalytic cracking, as the environmental conditions of the reservoir changed during geologic time (Levorsen 1954):

$$(\mathrm{Gas}_{Sp.Gr} = 0.0007;\ \mathrm{Oil}_{Sp.Gr} = 0.7 - 1.0;$$
$$\mathrm{Water}_{Sp.Gr} = 1.0 - 1.2).$$

1.9 Petroleum Province—Definition

A petroleum province is a region in which a number of oil and gas pools and fields occur in a similar geologic environment, e.g., the mid continent province of the south central United States has definite regional characteristics of stratigraphy, structure, and oil and gas occurrence; thus the term "province" has a specific meaning for geologists and the petroleum industry. Subprovinces may occur within provinces. The occurrence of petroleum is not confined to the land areas of the world; there are vast potential sources of petroleum underlying the shallow waters that border the continents. These submerged lands, sloping down to a depth of 600 ft (191 m) below sea level, are known as continental shelves.

References

Levorsen AI (1954) Geology of petroleum. W. H. Freeman, San Francisco, pp 540, 541, 574; Fig. 12.2, p 547
Gatlin C (1960) Petroleum engineering—drilling and well completion. Englewood, New Jersey, Prentice-Hall, p 19

Chapter 2

Sedimentary Basins and Plate Tectonics

2.1 Scope

The outer part of the earth's crust, the lithosphere, is constructed predominantly of sedimentary rocks of Mesozoic, Tertiary, and Quaternary deposits extending to a depth of about 12 km on land and only 1 km or less on the ocean floors.

The crust itself is the bedrock of the outer layer of the earth and is relatively thin, 20–40 km thick under the continents, composed mainly of acidic igneous rocks (granites, granodiorites, etc.); but is much thinner, about 8 km thick or less, under the oceans, where it is composed mainly of basaltic rock. The crust differs in composition and thickness depending on whether it underlies the continents or the oceans. The crust is floating generally on a **solid mantle** about 3,000 km in thickness, and is composed of very dense rocks surrounded by outer shell layers of the ultradense and highly viscous liquid rock of the asthenosphere. The **core**, which is the innermost solid iron mass, with a radius of about 3,500 km, is surrounded by the mantle.

During the past three decades (since 1967), the theory of plate tectonics has dominated geological thinking. According to this theory, the surface of the earth's crust (under both continents and oceans) is made up of rigid, aseismic lithospheric blocks which are slowly and intermittently in motion. There are major and minor plates, together with more complex smaller units. The plates are separated by linear "active bands," in which vulcanicity and seismic activity seem to be concentrated (Fig. 2.1; Komatina 2004).

The plate tectonic theory of continental displacement replaced the geosynclinal theory of the development of the folded mountain ranges of the earth's crust; the term "geosyncline" was originally used to describe

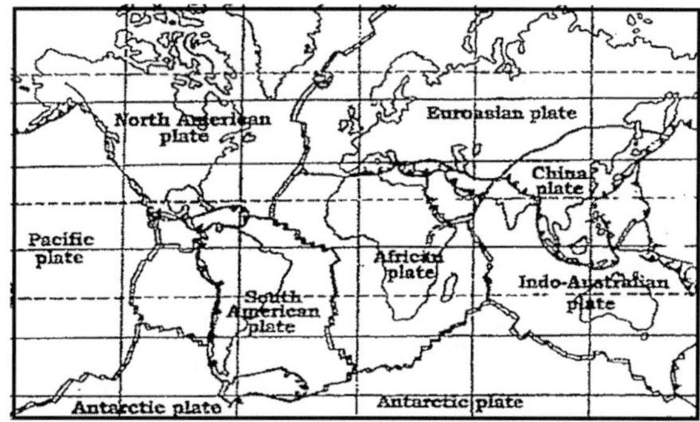

Fig. 2.1 Position of main tectonic plates (Komatina, 2004)

the long, narrow, subsiding depressions in which thick sediments accumulated and were affected by orogenesis (violent folding and uplift accompanied by volcanic activity). In general, the average rate of recent plate movements is usually from 1 to 2 mm/yr (Komatina 2004). There are two forms of tectonic movements:

(a) **Active tectonic movements** consist of epeirogenic movements (upliftings and downwellings) and orogenic movements (mountain formation on "geosynclinal" areas), which have led to Alpine and Mesozoic mountain ranges where tectonic activity was manifested in powerful seismicity and the development of volcanism, e.g., in the Alpine-Mediterranean zone in Eurasia.

(b) A **passive tectonic system** is characteristic of a large part of the Precambrian platforms, and applies to the entire territory of North American and Russia, and the large central zone of Africa, e.g., the Algerian Saharan platform, where the relief is level, low-lying in some areas and elevated in others.

F.A. Assaad, *Field Methods for Petroleum Geologists,*
© Springer-Verlag Berlin Heidelberg 2009

2.2 Continental Shelves

The earth's crust consists of two types, continental and oceanic, below which lie the vastly more bulky and heavier mantle and core. The continental crust, which is three to four times as thick as the oceanic crust, includes all the major land masses with submerged borders, known as continental shelves. The boundary zones between the continents and the oceans form the continental margins, which are made up of three components: (1) the continental rises, which largely comprise the "fans" of continental sediments; (2) the gently dipping shelves 5–250 miles in width, which constitute the submerged edges of the continents; and (3) the steeper slopes, about 10–30 miles wide, the bases of which mark the transition zone between the continental-type crust and the oceanic-type crust.

Under the 1958 Geneva Convention, the continental shelf is legally defined as "the sea bed and subsoil of the submarine areas adjacent to but outside the area of the territorial sea to a depth of 200 m or, beyond that limit, to where the depth of the superjacent waters admits the exploitation of the natural resources of such areas."

From the edge of the continental shelf, the surface of the continental crust slopes gradually downwards to the deep oceanic basins, which are underlain by the thinner oceanic-type crust. The thickness of the continental crust decreases gradually below the continental slope until the thinner oceanic crust is reached (Fig. 2.2).

Fig. 2.2–The world's continental shelves include the major seas, lakes, and some marine platforms within the 200-m depth range (Petroleum Press Service 1951).

In some oil field provinces, a broad "geosynclinal belt" of much sedimentation, which typically evolves to a large asymmetrical structural basin by being subjected to orogenesis, is the consequence of developments that have taken place over a long period of time. Sedimentary basins of such type are termed active when the floors have continued to sink and the depressions thus formed due to gravitational forces have filled with sediments. On the gentler side, older rocks underlying the sediments may outcrop on a part of one of the major shields where faulting is common in some belts, especially along the hinge line. On the more disturbed side (of the mobile belt side), there are foothills and mountain ranges; strong folds and overthrusting occur. The thickness of the sediments formed is less on the shelf or platform side than elsewhere.

Fig. 2.3–An asymmetrical structural basin "geosyncline trough" showing a maximum thickness that lies between the shelf side and the mobile rim where strong structures have developed; examples of this broad pattern are the Arabian shield of Saudi Arabia and the area of the Tigris-Euphrates valley of Iraq (Hobson 1975)

2.3 Plate Interactions

In plate tectonic theory, composite continents are assembled by crustal collisions that occur when the consumption of oceanic lithosphere beneath arc-trench systems results in the closure of an oceanic basin. The arrival of a continental block at a subduction zone where the intervening oceanic lithosphere was consumed, will thus throttle subduction, and the

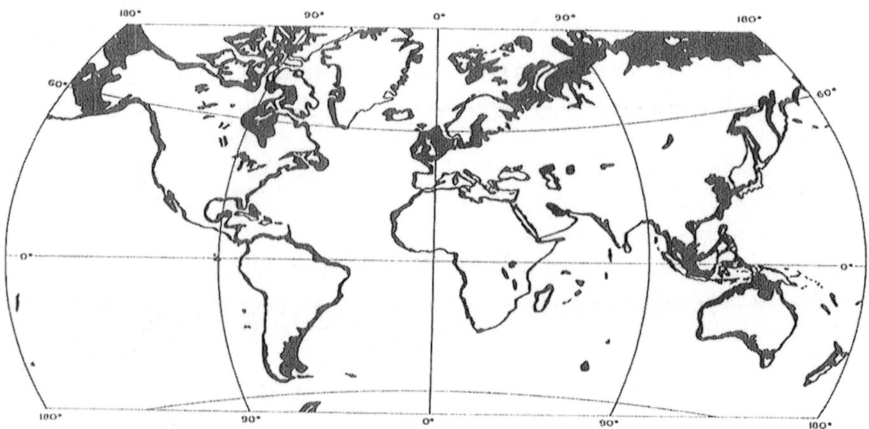

Fig. 2.2 World continental shelves (Petrol. Pres Service, 1951)

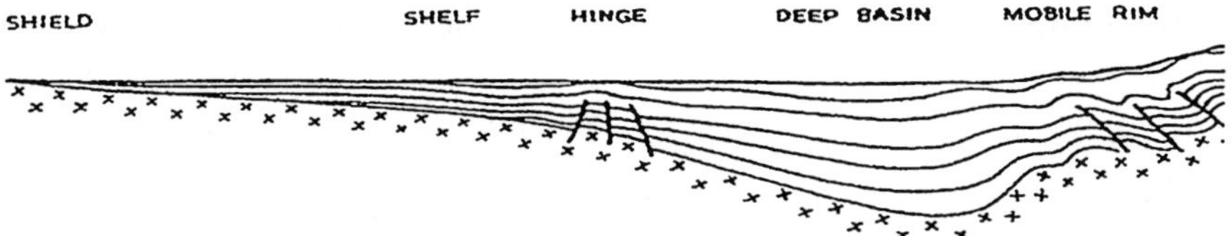

Fig. 2.3 A cross section of sedimentary basin its length of several hundred miles, and the thickness of the sediments overlying the basement could be a maximum of 10 miles. It lies between a stable shield (foreland) and a mobile belt. (Modified from Hobson, 1975)

position of the previous subduction zone will be taken by a crustal suture belt marking the line of tectonic juxtaposition of the two continental blocks involved in the crustal collision (Fig. 2.4). Examples of continental collisions in which one of the continental margins has a marginal arc-trench system and the other has a rifted-margin sediment prism (Dickinson 1974).

All plate interactions that involve construction of new lithosphere or consumption of old lithosphere as a result of large horizontal motions of plates, involve significant vertical motions of the lithosphere. There are three basic causes of subsidence or uplift as a result of plate interactions: (1) changes in crustal thickness; (2) thermal expansion or contraction of the lithosphere; and (3) broad flexure of plates of the lithosphere in response to local tectonic or sedimentary loading.

From a kinetic point of view, there are three kinds of plate junctures, analogous to the three classes of faults as defined by relative displacements:

(a) **Divergent plate junctures** (analogous to normal faults)—A separation of two plates (sea-floor spreading) occurs and causes rupture of the intact old lithosphere, which in turn results in intercontinental rifting when an incipient rift crosses a continental block. The rate of spreading may average several centimeters per year and might produce over geologic time the continental drift that represents the separation of Africa and South America and the associated growth of the Atlantic Ocean.

The plates diverge or move relatively apart and result in sea-floor spreading, in which magmatic material from the underlying lithosphere wells up in between to form linear lava ridges—parallel, linear "mid-ocean ridges"—which are found in the Atlantic, Pacific, and Indian Oceans.

(b) **Transform plate junctures** (analogous to strike-dip faults)—One plate slides laterally past the other along a transform or deep fault, without

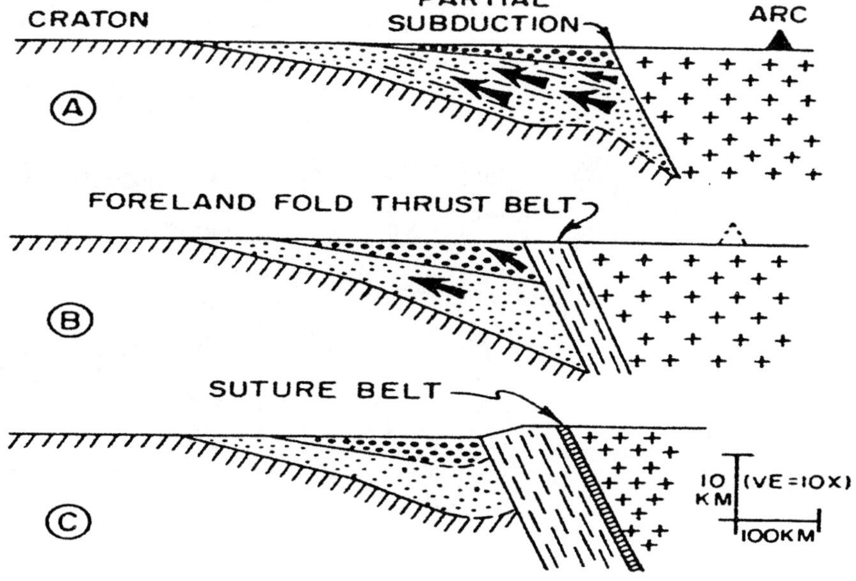

Fig. 2.4 Development (schematic) of foreland basin (heavy stipples) and suture belt from continental collision between arc-trench system (crosses) and rifted-margin sediment prism (light stipples). A, early stage of encounter; B, full collision; C, postcollision uplift of foreland fold-thrust belt. Heavy arrows show induced migration of hydrocarbons. (Dickinson, 1974)

accretion or consumption. However, hybrid plate boundaries also occur in some areas, as in oblique-slip faults, where some component of extensional or contractional motion occurs along a transform, and hence the two terms *transtension* and *transpression* are used to describe the interaction.

In accordance with the transform movements of the plates, the mid-ocean ridges in the Atlantic and Pacific Oceans, which run roughly north/south, are broken into a series of segments, each about 200 miles long, together with their related offsets, at the points of which are "transform fault zones," formed roughly at right angles to the mid-ocean ridges, and projecting above the deep ocean floors; these become earthquake and volcano zones.

(c) **Convergent plate junctures** (analogous to thrust faults)—When two oceanic plates collide, one plate is thrust at an angle beneath the other, and dives down into the mantle where it is partially destroyed by heat. Convergent junctures are sites of plate consumption where oceanic lithosphere formed previously at a divergent plate juncture, descends into the mantle. Part of the slab continues to sink to a depth of about 700 km, where it comes to rest; the part which rises to merge with the upper plate is converted to low-density magma.

Fig. 2.5–A diagram of a subduction of an oceanic plate, with evolution of volcanic island and trench (Hobson 1975)

Fig. 2.6–Two principal kinds of plate interactions at divergent and convergent plate junctures (Dickinson 1981)

Hobson (1975) discussed the convergence of plates according to a different approach, depending on whether oceanic- or continental-type plates are involved.

(1) A **coastal mountain range** is formed along the leading edge of the continental plate, when it is advancing towards and overriding a relatively stationary oceanic plate.

(2) **Island arcs and subduction zone trenches** are formed when an oceanic plate is advancing and passing beneath a relatively stationary continental plate.

(3) **Major mountain ranges** are formed when two continental-type plates slowly converge, as a result of the squeezing of the sediments carried on the underthrust plate (e.g., the Himalayas in India).

2.4 Development of Sedimentary Basins

Plate tectonics theory is generally related to lateral and vertical motions of plates that form sedimentary basins. A sedimentary basin develops as an accumulated prism of strata, resulting either from subsidence of the basin floor, or from uplift of confining basin margins. Each basin has its own unique history connected to a particular sequence and combination of plate interactions and depositional conditions. During basin evolution, the stratigraphic fill of the basin took place because of the activity of the depositional systems. Folded and faulted structures within the basin were formed because of either tectonic or sedimentary

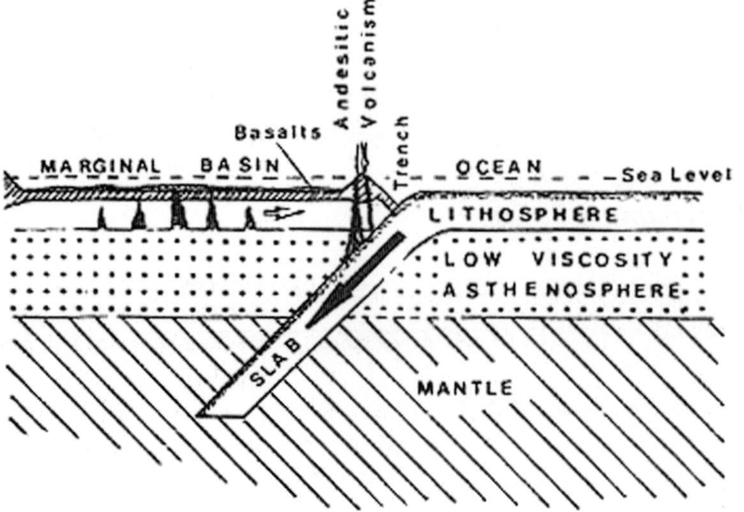

Fig. 2.5 Diagram illustrating subduction of an oceanic plate, with evolution of volcanic island arc, marginal basin and trench. (after Sleep and Toksoz. Nature. 547, Oct. 22, 1971, in Hobson, 1975)

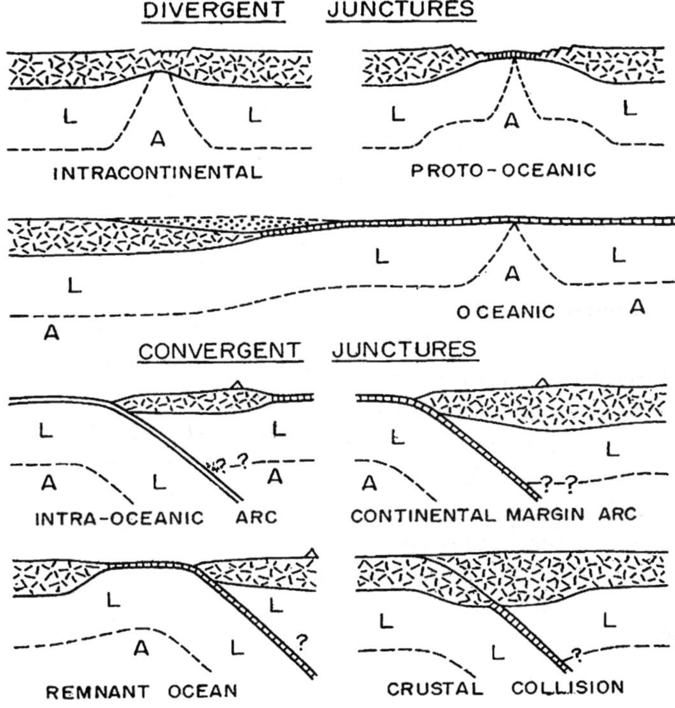

Fig. 2.6 Principal kinds of plate interactions at divergent (above) and convergent (below) place junctures showing relations of lithosphere (L) and asthenosphere (A) as well as crust (ornamented) (Dickinson, 1981)

evolution. Extensional deformation usually produces normal faults and tilted blocks, whereas contractional deformation produces folds and thrust faults.

Plate movements have lately been considered to result from the presence of "hot spots," which represent a number of thermal centers, and are fixed in the upper mantle, from which "plumes" of hot material rise intermittently to burn holes in the overlying crust. Consequently, the continental plates are pushed away from these "hot spots" by the creation of new ocean floor; and when the movement of the lithosphere above the plume occurs by the process of sea-floor spreading, a "plume scar" is left on the crust in the form of a line of volcanic cones.

2.4.1 Geodynamic Settings of Sedimentary Basins

Sedimentary basins can occur in two general kinds of geodynamic settings.

(1) **Rifted settings occur** where divergent plate motions and extensional structures dominate and are later enhanced by thermal decay as time passes and may be augmented mostly by flexures in response to sedimentary loading.

(2) **Orogenic settings** occur initially where convergent plate motions and contractional structures dominate; subsidence occurs initially by plate flexures, either because of plate consumption or because of local tectonic thickening of crustal profiles, possibly augmented by sedimentary loading and other thermotectonic effects. The lithosphere may exchange locations between rifted and orogenic settings, and hence the name composite basins, in terms of plate tectonic settings. Basins of rifted settings, in which tectonic evolution is dominated by extensional plate motions and crustal rifting, and those of orogenic settings, in which tectonic evolution is dominated by contractional plate motions and orogenic deformation, include several idealized subgroups.

Oceanic basins evolve regularly from nascent phases dominated by extensional tectonics into remnant phases dominated by contractional tectonics. Similarly, the sedimentary assemblages along rifted continental margins include phases deposited within protoceanic rifts underlying the younger miogeoclinal prism, which may in turn be covered and flanked by the

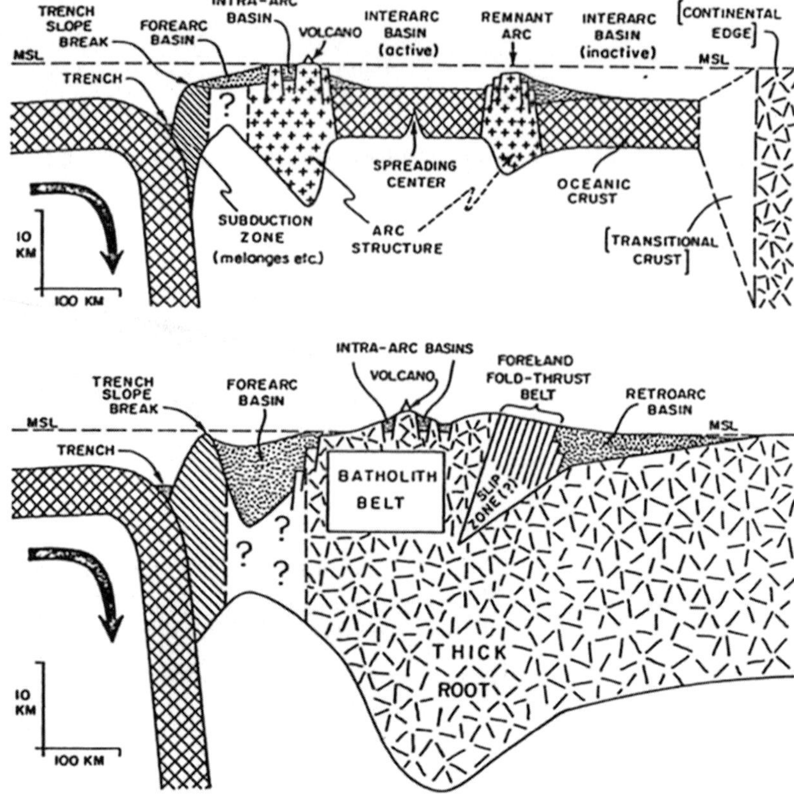

Fig. 2.7 Schematic diagrams (vertical exaggeration 10X) to illustrate sedimentary basins associated with intraoceanic (above) and continental margin (below) magnatic arcs. (Dickinson, 1981)

deposits of a progradational continental embankment. These rifted-margin sedimentary associations may later be covered in part, with the onset of orogeny, by the foreland deposits of retroarc, or peripheral basins as arc (Fig. 2.7) or as collision (Fig. 2.8, Dickinson 1981).

2.4.2 Sedimentary Basins and Hydrocarbon Occurrences

Dickinson (1981) discussed the accumulation of oil and gas within the thick prisms of sediment deposited along many rifted continental margins. The plates apparently rest upon a mobile layer called the asthenosphere, within which the geothermal gradient across the plates is controlled mainly by the conductivity between the basal temperature (that of the asthenosphere) and the surficial temperature (that of the troposphere [atmosphere and hydrosphere]). The rigid blocks of plates are spherical caps, or arcuate slabs, of the earth's lithosphere that extend down to the so-called low velocity zone of the upper mantle. The amount, nature,

and location of fluid hydrocarbons within the basin is governed by the geometric shape and size of the basin, the nature of the stratigraphic fill and the types of structures, and most importantly by the thermal history of the basin.

Analysis of the plate tectonics of sedimentary basins should identify styles and times of basin evolution and trends that embody favorable combinations of the following four important attributes for hydrocarbon occurrence. (1) Organic-rich source beds within the sedimentary sequence are generally fine-grained sediments deposited in black-bottom areas where oxidation by aerobic decay during early diagenesis is limited, and rapid burial enhances the likelihood of preservation. Basins in which sedimentary fill includes the favorable sites that harbor the largest sources of potential hydrocarbons exist in oxygen-minimum zones on open slopes and in low-oxygen zones within saline nonmarine or marginal marine environments. (2) Fluid hydrocarbons are normally generated by appropriated heat for thermal maturation of liquid hydrocarbons or thermal gas. (3) Permeable migration paths for the concentration of fluid hydrocarbons should have access

Fig. 2.8 Schematic diagrams (vertical exaggeration 10X) to illustrate sedimentary basins associated with crustal collision to form intercontinental suture belt with collision orogen. Symbols: TR, trench; SZ, subduction zone; FAB, forearc basin; RAB, retroarc basin; FTB, foreland fold-thrust belt. Diagrams A-B-C represent a sequence of events in time at one place along a collision orogen marked by diachronous closure; hence, erosion in one segment (C) of the orogen where the sutured intercontinental join is complete could disperse sediment longitudinally past a migrating tectonic transition point (B) to feed subsea turbidite fans of flysch in a remnant ocean basin (A) along tectonic strike. (Dickinson, 1981)

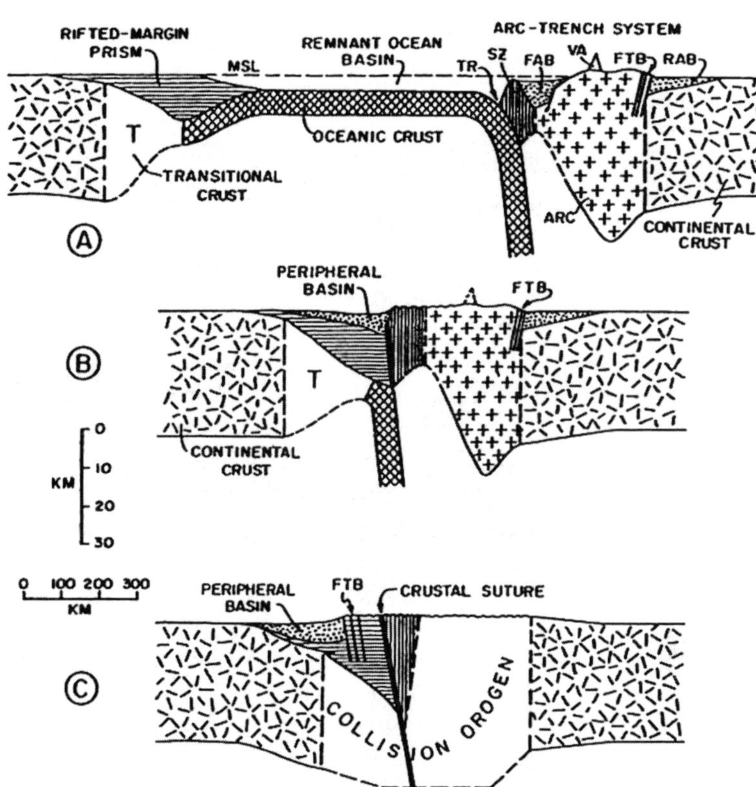

to gather any hydrocarbons produced. The most effective carriers are tilted conduit beds of well-sorted sandy strata contained beneath impermeable sealing beds of shales or eruptive rocks. (4) Petroleum and petroleum hydrocarbons can be contained in porous reservoir beds (whether clastic or carbonate), and are confined by impermeable capping beds or traps that may be formed either by stratigraphic enclosures or by structural features of tectonic or sedimentary origin.

Conceivably, plate tectonics may lead to a comprehensive theory of hydrocarbon genesis. Plate interactions fully explain the causes of subsidence, the sequencing of depositional events, the development of structural features, and the timing of thermal flux—all of which in turn can explain conditions in the geologic record of sedimentary basins (Dickenson 1981).

prisms are drawn down against and beneath the suture belts formed by crustal collisions that assemble composite continents. During such partial subduction, fluid hydrocarbons may be driven updip away from the subduction zones to accumulate in reservoirs along adjacent platform margins and within growing foreland fold-thrust belts, e.g., the immense petroleum accumulations of the Persian Gulf southwest of the Zagros suture belt. Similar oil migration in response to partial subduction may have influenced the distribution of petroleum in foreland basins in general; if so, strategies for exploration can perhaps be improved by taking this factor into account. Dickinson (1974) assumed that there were possible direct relations between plate movements accompanying subduction and the opening of migration paths for petroleum.

2.5 Oil Migration and Subduction

Petroleum and natural gas pools are abundant within the thick prisms of sediment deposited along many rifted continental margins. The rifted-margin sediment

2.6 Rifted-Margin Sediment Prisms

In plate tectonic theory, ocean basins are initiated by continental separations that begin with **intracontinental rifting**. When complete separation of continental fragments is achieved, the raw edges of continental blocks along rifted continental margins are in preferred positions to receive thick prisms of sediment, composed both of debris washed off the adjacent continental blocks and of biogenic calcareous materials that were deposited in shoal areas at the flanks of the adjacent oceans.

The growing rifted-margin sediment prism imposes sufficient load on the lithosphere near the interface between old continental and new oceanic crust to induce downbowing of the lithosphere by flexure (Fig. 2.9, Walcott 1972). The flexure tilts the surface of the continental block seaward to allow landward parts of the rifted-margin sediment prism to encroach as much as 100–250 km beyond the initial continental edge, which may be depressed as much as 2.5–5 km. Continued growth of the rifted-margin sediment prism may in time allow its oceanward edge to advance well into the adjacent oceanic basin until the continental shelf, and perhaps even part of an accretionary coastal plain, come to stand fully above the oceanic basement. The type of rifted-margin sediment prism developed by wholesale sedimentary progradation of the continental edge has been aptly termed a continental embankment (Dietz 1963).

Accumulations of petroleum in rifted-margin sediment prisms are well known. The natural association of offshore source beds and nearshore reservoir beds of various kinds is inherent. The following two factors combine to exert a pumping action that drives fluid hydrocarbons updip from offshore source beds into favorable reservoir beds. (1) The **progressive loading** of offshore source beds because of depositional growth of the rifted-margin prism; and (2) The **progressive seaward tilting** of the continental edge and the successive layers of the rifted-margin prism because of flexural bending of the lithosphere under the growing sedimentary load offshore. These factors are probably most influential within fully developed continental embankments, where potential conduit beds, intercalated with sealing beds, connect offshore source beds with inshore facies suitable as reservoir beds without interruption by thin slope facies (Dickenson 1974).

In conclusion, the largest reserves of petroleum are inferred to accumulate where conditions are most favorable for long-distance updip migration of oil from offshore source beds into attractive reservoirs in nearshore deposits along or near platform margins. The requisite regional dips and overburden loads are attained within the sediment prisms deposited along rifted continental margins, and also beneath foreland basins associated with foreland fold-thrust belts adjacent to major orogens. Apparently, optimum relations for oil concentration thus occur in the Middle East where a rifted-margin sediment prism was drawn partly into a subduction zone, now a suture belt marking the site of continental collision, and was partly buried beneath a foreland basin sequence (Dickenson 1979).

See Glossary—Plate tectonics—Definitions; and examples of plate junctures.

Appendix 2-A.1

Davis and Elderfield (2004) submitted a concise review of the relatively brief history of hydrogeology, which began shortly after the revival of plate tectonic theory little more than thirty years ago. It describes the nature and important consequences of fluid flow in the sub-seafloor, ending with a summary of how the oceans are affected by the surprisingly rapid exchange of water between the crust and the water column overhead.

The importance of fluid circulation below the seafloor and the exchange of water between the crust and the oceans can be easily appreciated by considering that the oceanic crust constitutes the most extensive geological formation on earth, and that hydrologic activity within it extends from mid-ocean ridges to subduction-zone accretionary prisms. The

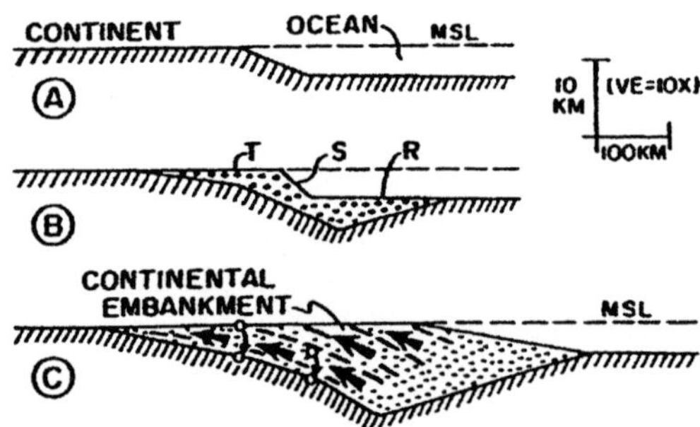

Fig. 2.9 Growth (schematic) of rifted-margin sediment prism (stippled), A, rifted continental margin without sedimentation; B, continental terrace (T)-slope (S)-rise (R configuration; C, progradational continental embankment; small arrows show rotational tilt of basement along continental margin downward toward ocean basin; heavy arrows show updip migration of hydrocarbons parallel to bedding (dashes). (Walcott, 1972)

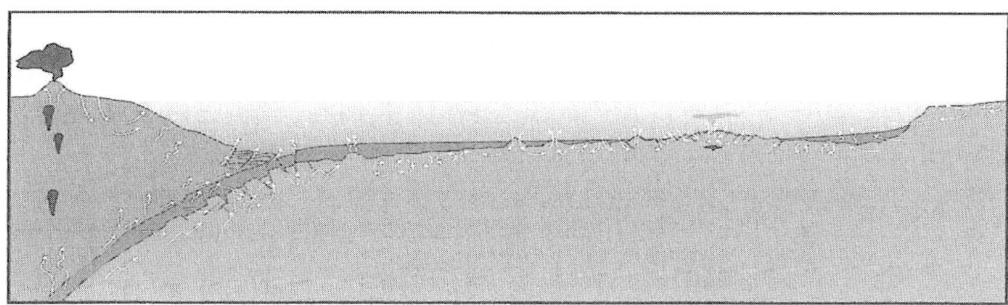

Fig. 2.10 Schematic cross-section depicting various types of sub-seafloor fluid flow, ranging from topographically driven flow through continental margins and consolidation-driven flow at subduction zones to thermal buoyancy driven flow at mid-ocean ridge axes and in the oceanic crust beneath broad regions of the oceans. Oceanic igneous crust and sediments are shown light gray and darker gray, respectively, and magma is shown in black: flow is depicted by white arrows. The figure was originally prepared for the Itegrated Ocean Drilling Program Initial Science Plan. 2003–2013. and is reproduced courtesy of JOI, Inc. (Devis and Elderfield, 2004)

upper part of the crust is characterized by very high permeability, and it is host to huge fluxes of water. At the ridge axes occur the most spectacular and directly observable manifestations of flow; and at the seafloor, heat from magmatic intrusions drives high-temperature springs (Fig. 2.10).

References

Komatina MM (2004) Medical geology—effects of geological environments on human health. Fig 2.12; ISBN 0-444-51615-8, Elsevier; USA, www.elsevier.com

Dickinson WR (1981, 1974) Subduction and oil migration. Geology 2/1974, pp 1519–1540, 1950

Dietz RS (1963) Wave base, marine profile of equilibrium, and wave-built terraces: a critical appraisal. Geol Soc Am Bull 74:971–990, *in* Dickinson (1974)

Davis E, Elderfield H (2004) Hydrogeology of the oceanic lithosphere. Cambridge University Press, New York, NY, Fig 1; publicity@Cambridge.org

Hobson GD (1975) Introduction to petroleum geology—modern tectoinc theories. three para: pp. 240, 241; Figs 69 & 70/p242; Scientific Press Ltd., Beaconsfield, England

Walcott RI (1972) Gravity, flexure, and the growth of sedimentary basin at a continental edge. Geol Soc Am Bull 83:1845–1848, *in* Dickenson (1974)

Falcon NL (1973) Exploring for oil and gas. *in* Modern petroleum technology, 4th ed. Hobson GD and Pohl W eds (1973)

Chapter 3

Surface Geophysical Petroleum Exploration Methods

3.1 Introduction

Surface geophysical techniques determine density, magnetic, and acoustical properties of a geologic medium. Three geophysical methods used in petroleum exploration comprise magnetic, gravimetric, and seismic (including refraction/reflection) techniques. The magnetic and gravity methods are used only in primary surveys where little is known of the subsurface geology and/or the thickness of sediments of potential prospective interest. The seismic reflection method is universally used for determining the underground geological structure of a reservoir rock in a certain area. The method(s) selected will depend on the type of information needed, the nature of the subsurface materials, and the cultural interference.

3.2 Magnetic Survey

A magnetic survey is primarily used to explore for oil and minerals. Magnetic exploration is based on the fact that the earth acts as a magnet. Any magnetic material placed in an external field will have magnetic poles induced upon its surface. The induced magnetization (sometimes called polarization) is in the direction of the applied field, and its strength is proportional to the strength of that field. The location of an area in relation to the magnetic poles is measured by the inclination of the earth's field or the "magnetic inclination" (USGS, U.S. Army, 1998).

Aeromagnetic Surveys were developed in wartime to overcome the problem of detecting submerged submarines from aircraft; they have since gained considerable success in petroleum and mineral exploration. Both the aeromagnetic surveys and the airborne radioactivity surveys can be classified as "remote sensing techniques." The advantages of the use of airborne surveys over ground instruments are the speed of the surveys and the possibility of reaching otherwise inaccessible areas.

3.3 Gravimetric Survey

Variations in gravity depend upon lateral changes in the density of earth materials in the vicinity of the measuring point. Many types of rocks have characteristic ranges of densities, which may differ from other types that are laterally adjacent. Driscoll (1986) stated that the earth's gravitational attraction at a particular site is a function of the density of the surface sediments and the underlying rock units. The density variations may be attributed to changes in rock type (porosity or grain density), degree of saturation, fault zones, and the varying thickness of unconsolidated sediments overlying the bedrock. Thus an anomaly in the earth's gravitational attraction can be related to a buried geological feature, e.g., a salt dome or other deposit which has limited horizontal extent. Actually, all geophysical surveys concentrate on the discovery of "anomalies" in the rocks which overlie or surround possible petroleum accumulation.

F.A. Assaad, *Field Methods for Petroleum Geologists*,
© Springer-Verlag Berlin Heidelberg 2009

3.4 Seismic Exploration Survey

3.4.1 General

The word "seismic" refers to vibrations of the earth, including both earthquakes and artificially created sound waves that penetrate into the earth. Sounds measured are in the frequency range of about 10–100 cycles s^{-1}. The depths investigated for a sound to travel into the earth and return are as much as 16 km.

Seismic investigations depend on the fact that elastic waves (or seismic waves) travel with different velocities in different rocks. It is possible to determine the velocity distribution and locate subsurface interfaces where the waves are reflected or refracted by generating seismic waves at a given point and observing the times of arrival of those waves at a number of other points (or stations) on the surface of the earth.

Seismic surveys which are based on the velocity distribution of artificially generated seismic waves in the ground are produced by hammering on a metal plate, by dropping a heavy ball, or by using explosives. Energy from these sources is transmitted through the ground by elastic waves, which are so called because, when the waves pass a given point in the rock, the particles are momentarily displaced or disturbed, but immediately return to their original position or shape after the wave passes.

3.4.2 Seismic Refraction Methods

Seismic refraction methods were originally developed to locate concealed masses of salt plugs (e.g., in Algieria, Mexico, and Germany), and to trace major anticlinal axes in massive limestones (e.g., in Iran) which could not be located with surface geology. In both cases, the resulting shock waves caused by the explosives travel faster through salt and limestone than through the associated sedimentary rocks. Seismic refraction is also a useful reconnaissance tool for determining the depth of a high velocity metamorphic or igneous basement below a small sedimentary basin, etc.; each geologic formation has a characteristic seismic velocity that affects the arrival time.

3.4.3 Seismic Reflection Method

The seismic reflection method depends on the echo sounding principle and is a special tool for oil and gas exploration; it records reflected shock waves from a number of successive beds and their angle of inclination along the line of observation. The method uses a seismic wave produced by a weight dropping, a hammer blow, or another seismic source that is reflected off the bedrock and returns directly to the geophone, where the elapsed time is recorded. Hammer stations are usually at 9.1 m or less from the geophone to maximize the reliability of the reflected wave energy. The operator strikes a hammer plate at five to ten sites that are within 9.1 m of the geophone. The seismic signals received from these sites are summed automatically by the seismograph, canceling out the surface waves and other extraneous impulses, and the primary reflected wave is prominently displayed on the cathode ray tube (Driscoll 1986).

3.4.3.1 Multiple Reflections and Marine Exploration

Special ships allow rapid surveying in marine exploration where the presence of water as the medium for inducing the shock waves give remarkably good results. A system known as vibroseis employs a non-impulsive sound source with transmitters like huge loudspeakers with their diaphragms pressed against the ground. Because the seismic signal is quite unrecognizable until it is "pulse-compressed" by the correlation process, the vibroseis input can be regarded as a degenerate form of seismic impulse. Degeneration of a seismic impulse often occurs naturally in its travel through the earth, the commonest form being reverberation in a "ringing layer." Ringing is often a big problem in marine exploration. The shot has two very good reflecting interfaces immediately above and below it, namely, the sea surface and the sea bed. The downward traveling impulse from the shot is followed closely by a reflection from the sea surface, and when reaching the sea bed, it is partially reflected back to the surface, only to be once again sent on a downward course. Therefore the initial downward wave has three "ghosts" following it, leading to an endless process; this process continues as part of the more general problem of multiple reflections (Tarrant 1973).

3.4.3.2 Seismic Profiling of Diapiric Structures

The reflecting seismic process records reflections from acoustic discontinuities in the subsurface by generating a sound wave near the surface and detecting the reflected energy return from subsurface discontinuities.

Features associated with salt domes identified by seismic sections are radial faulting, the doming of overlying strata, the dip of sediments on the flank, nonconformity and wedging effects, and the development of rim synclines. A modeling process permits accurate interpretation of these features, e.g., piercement diapirs, salt dome structures at different stages of development, and associated geologic phenomena, thus increasing success in drilling for hydrocarbon reserves.

3.5 Land-Satellite Images in Salt Dome Exploration

Evaporites, including salt domes, are formed by the evaporation of brines because of dryness in arid conditions; they are normally interbedded with carbonate rocks together with red and green shales in cyclical sequences. In some parts of the world, buried evaporite beds lie several hundreds of meters beneath the ground surface, and have generated salt plugs or salt domes which have moved upwards through the overlying beds and probably appear on the surface in "diapiric" or piercing plastic flow, e.g., the Zechstein of North Germany, the Triassic of the Gulf region and Algieria, as well as the Miocene of the Suez Canal of Egypt.

Normal aerial photographs can be used to detect certain geological and ecological features peripheral to many ore deposits or oil and gas fields. Landsat images can be used by adapting remote sensing methods to the images; early space imagery displaced extended structural elements such as closed anticlines, domes, intrusive bodies, folded mountain belts, fault zones, and regional joint patterns.

Remote sensing data obtained from satellite images are most beneficial to proper and more accurate interpretations of the earth's surface. The use of aircraft to obtain data in locating a resource target considerably reduces the cost of exploration, but the use of spacecraft to obtain remote sensing information reduces the overall cost of the ground survey even further; and therefore "the higher we go, the deeper we can see" (Trollinger 1968).

References

Driscoll FG (1986) Groundwater and wells. Johnson Division, UOP Inc., St. Paul, Minnesota 55165, 3-para/ pp. 168–177.

Tarrant LH (1973) Geophysical methods used in prospecting for oil, *in* Hobson GD and Pohl W eds. (1973) Modern petroleum technology, p. 81, Applied Science Publishers Ltd; England: The Institute of Petroleum Geology.

Trollinger WV (1968) Surface evidence of deep structure in the Delware Basin: "Delware Basin Exploration", Guidebook. West Texas Geol. Soc. Pub. #68-55, pp. 87–104; *in* Halbouty MT (1967) Salt domes.

USGS, U.S. Army (1998) Earth Science Applications. National Training Center, Fort Irwin, California (http://wrgis.wr.usgs.gov./docs/geologic)

Chapter 4

Drilling Technology in Petroleum Geology

4.1 Introduction

There are two drilling methods used in the petroleum industry: (1) The **cable tool method**, by which the first oil well was drilled in 1859 to a depth of 65 ft, was first employed by the early Chinese in the drilling of brine wells. (2) The **rotary drilling method**, started by a French civil engineer in 1863, is the most common method that performs a rotary grinding action; some cable tool rigs are still working in parts of Europe as well as in the USA.

Rotary drilling methods are much more effective in drilling shallow, unconsolidated sands than the cable tool operations.

A cable tool rig is a percussion drilling apparatus consisting of the drill string, which is mainly composed of the drill bit, made of a heavy steel bar; the drill stem, a cylindrical steel bar screwed directly above the bit; jars of heavy steel links, to produce a sharp upward blow on the tools; and tool joints that connect these parts.

In a standard cable tool rig, there are three rig lines or cables: the drilling line, the sand line, and the calf or casing line (used to run the casing into the well). The derrick floor supports the drilling line, the sand line on which the bailer is run, and the casing line (Brantly 1952; Fig. 4.1).

The cable tool rig can be used to drill wells up to about 170 m, but it is generally confined to much shallower operations. In cable tool drilling, the hole is kept partly filled with water to soften the formation and prevent caving. At depth, when the bottom of the hole becomes full of cuttings, the rock bit should be withdrawn and a bailer run to remove the cuttings.

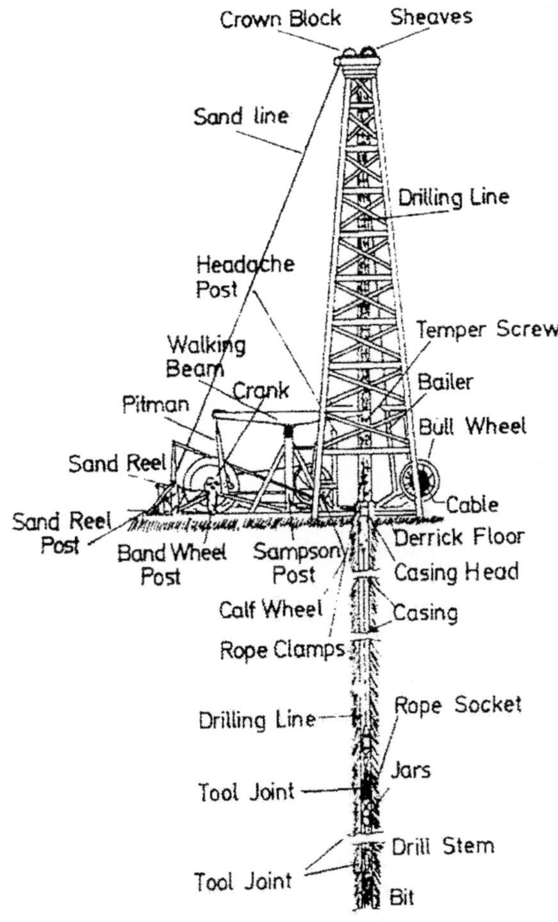

Fig. 4.1 Standard cable tool drilling system (Brantly, courtesy AIME)

A **rotary drilling rig** (API 1954; Fig. 4.2) comprises six components: (1) The **derrick and substructure**. The derrick provides the vertical clearance

F.A. Assaad, *Field Methods for Petroleum Geologists*,
© Springer-Verlag Berlin Heidelberg 2009

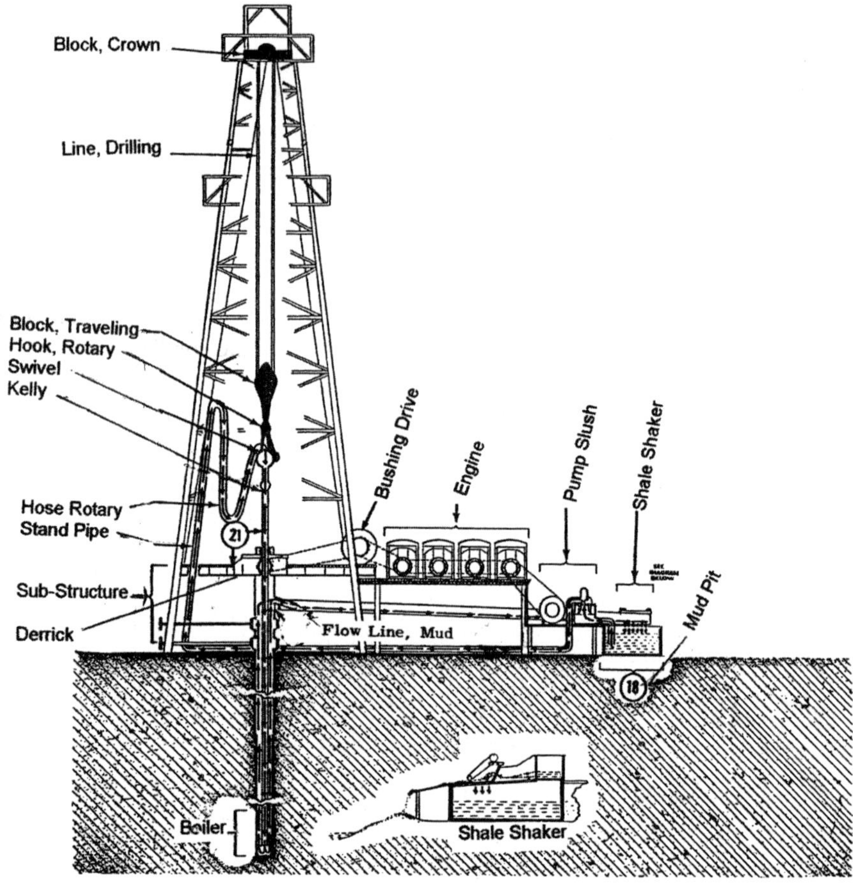

Fig. 4.2 Basic components of a rotary drilling rig. Courtesy API (1954)

necessary for raising and lowering the drill string into and out of the borehole during drilling operations, whereas the substructure is the support on which the derrick rests. (2) The **Mud Pump** circulates the drilling fluid at the desired pressure and volume. (3) The **drawwork** or **hoist** is the control center from which the driller operates the rig. It houses the drum and is the key piece of equipment on a rotary rig; it includes the clutches, chains, sprockets, engine throttles, etc. (4) The **Rotary Drill String** consists of the Kelly joint, drill pipe, tool joints, and drill collars. The drill pipe furnishes the necessary length for the drill string and serves as a conduit for the drilling fluid. Between the drill pipe and the rock bit are the drill collars, which are heavy-walled steel tubes to furnish the compressive load on the bit, thus allowing the lighter drill pipe to remain in tension. The drill string is an extremely expensive rig component and must be replaced periodically to avoid failure due to material fatigue that may result from corrosion and/or improper care and handling; The Kelly joint is the topmost joint in the drill pipe and is commonly square, hexagonal, or even octagonal. It passes through snugly fitting, properly shaped bushings in the rotary table, allowing the table's rotation to be transmitted to the entire drill string. (5) **Rock bits** are designed to give optimum performance in various formation types. (6) The **Drilling Line** affords a means of handling the loads suspended from the hook during all drilling operations.

In rotary drilling, the well is drilled by a rotating bit to which a downward force is applied; the bit is fastened to and rotated by a drill string, which is composed of high quality drill pipe and drill collars, with joints added as drilling progresses. The cuttings are lifted from the well by the drilling mud fluid, which is continuously circulated down, inside the drill string through the nozzles in the bit, and upward in the annular space between the drill pipe and the borehole.

The returning mud fluid discharged at the surface enters into a segregated sedimentation tank or pit that affords a sufficient period to allow cuttings—separation and any necessary treating—then the mud is picked up by the pump suction to repeat the cycle.

Turbo drilling of high speed rotation is used in Europe to avoid the rotation of a long, slim drilling column and its removal from the hole when the cutting tool is worn out. By using the turbo drill, the mud should be kept in good condition and free of sand or any other abrasive particles.

4.2 Petroleum Drilling Operations

4.2.1 Discussion

Petroleum wells are considered vertical wells in rotary drilling operations, although in practice borehole deviation exists and 3°–5° is commonly specified as the maximum acceptable deviation in vertical boreholes; in the early days, some wells happened to run into each other during drilling. In drilling operations after the development of reliable directional surveying equipment, the bottom hole location of many of the original wells deviated from the surface location; accordingly large reserves of recoverable oil were bypassed.

Subsurface mapping relying on depth measurements was found impossible to depend on; depths obtained from drill pipe measurements were quite misleading because of the deviation of wells from the verticality. It was not until 1950, when a later fundamental treatment of hole deviation was carried out successfully, that principal consideration was given to the angle between the hole and the vertical; whereas the compass direction of deviation was of secondary importance.

Directional drilling has recently been practiced to control angle deviation from the vertical in a borehole. This can be accomplished by Totco instruments that induce the deviation, followed by survey instruments that record the amount of deviation from the vertical and the direction of deviation. This process is applied in oil or gas reservoirs, when it is desirable to drill several wells from one location because of surface climatic conditions, or to extinguish an oil fire in a well; it is mainly used to avoid placing the bottom of a borehole at an inaccessible surface location where a vertical well site is impractical or impossible because of topographical and/or legal problems (Fig. 4.3; Eastman Oil Well Survey Co. 1960). Besides, the required bottom of the drilled hole may be under a lake or river, just off the coast, under a roadway, or in an area where legal access to the ideal location is barred. A number of specialized tools, commonly called primary deflection tools, are used in directional drilling for initiating and maintaining the desired borehole direction; but they are used as infrequently as possible because of the cost and rig time involved. Directional drilling requires the measurement of both vertical and horizontal directions, accomplished with various devices.

One device used to effect deviation is known as a **whipstock**—a tapered steel wedge with a concave groove on its inclined face; it guides the bit away from the previous course of the well towards the direction that faces the inclined groove.

Figure 4.4 shows directional drilling tools on salt domes using modern whipstocks and surveying instruments which can be grouped into three basic deflection patterns or types, depending on the geologic

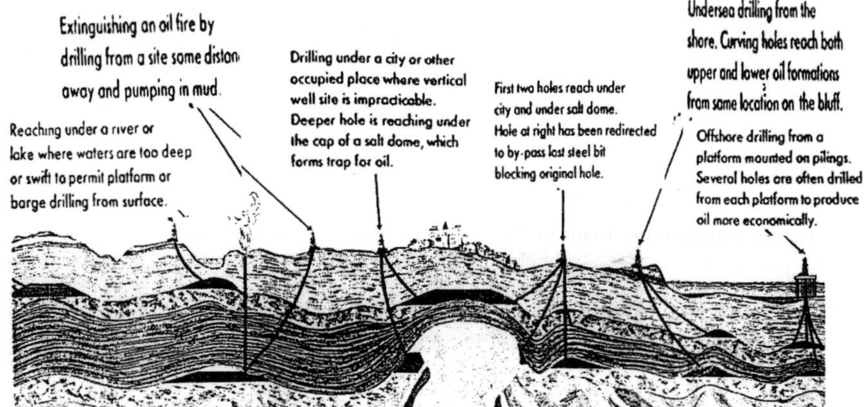

Fig. 4.3 Specific applications of directional drilling. Courtesy Eastman Oil Well Survey Company

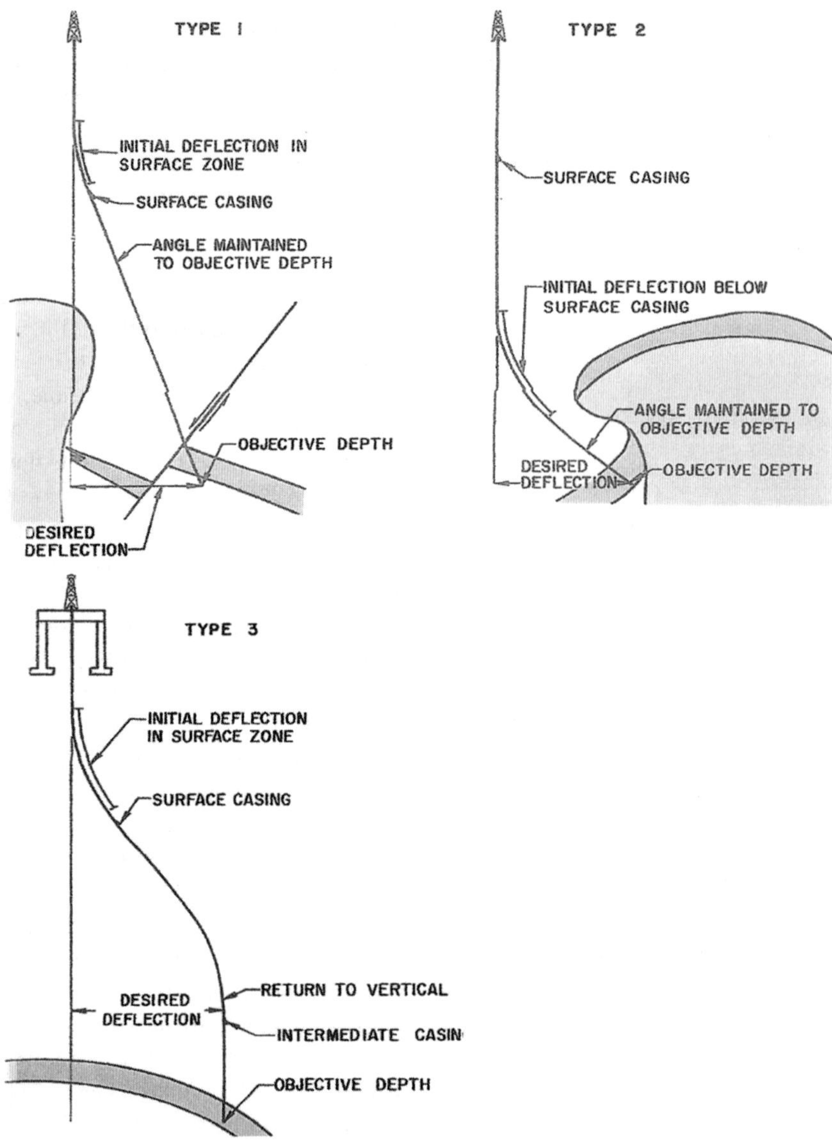

Fig. 4.4 Diagrams showing major types of directional drilling methods. Type 1 – Initial deflection in the surface zone with the deviation angle maintained to the objective depth. Used in shallow to moderately deep wells where small lateral displacement is required and where intermediate casing is not anticipated. Type 2 – Initial deflection below surface casing. Used for show deflections in multi-pay fields where multiple completions are not desired and to determine the extent and structural attitude of the reservoir. Type 3 – Initial deflection in surface zone and reduction of the angle of deflection to near vertical after desired horizontal deflection is achieved. Used in wells where intermediate casing is set to protect against problem zones and to achieve uniform bottomhole spacing (modified after Cook, 1957)

factors, the economics, and the desired final results (Halbouty 1967, modified from Cook 1957).

4.2.2 Types of Drilling Operations

1. Offshore Drilling—The basic drilling operations in both onland and offshore wells are exactly the same. The only difference is in the procedure of connecting the well to the rig, which is either floating or standing above the location. There are four basic forms of offshore rigs: fixed multiwell platforms, mobile jack-up platforms, drill-ships, and semisubmersible platforms. In offshore drilling operations, sea water is economically used as the base fluid for the salt-mud system. Seawater mud is effective in drilling shallow offshore domes; however, additional treating agents are necessary to overcome contamination by calcium and magnesium ions.

2. **Coring Techniques**—Core drilling is necessary to obtain and examine larger, unbroken pieces of reservoir rock samples which are obtained either from the bottom during drilling or from the side of the borehole wall after drilling.

   (a) **Bottom coring**, at the time of drilling, utilizes an open center bit that cuts a doughnut-shaped hole, leaving a cylindrical plug or core in the center which rises inside the core barrel and is then captured and raised to the surface for analysis. Diamond core bits are preferable; although they cost much more, they drill more total footage and when worn out, can be returned to their supplier for salvage.

   (b) **Sidewall coring after drilling** is often desirable to obtain core samples from a particular zone or zones already drilled. A special device is used which provides a hollow bullet that imbeds in the formation wall and is then fired from an electrical control panel at the surface. Samples of this type, are normally $\frac{3}{4}$ or 1 13/16 inch in diameter, and $\frac{3}{4}$ to 1 inch long. Sidewall coring is widely applied in soft rock areas, where hole conditions are not conductive to drill stem testing (DST); geologists can depend on electric logging for selecting zones to be sampled.

4.3 Drilling Fluids

4.3.1 Properties of Drilling Muds

Drilling mud properties vary with geologic conditions, since the degree of change in mud properties depends on the nature of the geologic beds being penetrated. For successful well completion in soft rock drilling areas, precise control of mud properties requires using expensive and complicated chemical mixtures. In hard rock drilling operations, plain water, air, and gas may replace the drilling fluids (see API Recommended Practices No. 29 1949).

When only "clean" water is circulated in the borehole, the water can pick up clay and silt and form a natural drilling mud. During this process, both the weight and viscosity of the drilling fluid increases. It is possible to attain a maximum weight of approximately 1.32 kg/l per gallon when drilling in natural clays.

When the weight of the drilling fluid or the hydrostatic pressure in the borehole exceeds that of the reservoir rock, fluid moves from the borehole into the lower pressure zone, where the incorporated fine particulate matter can be deposited during the drilling operation and infiltrated into the pore space of the zone, together with any solids added to the drilling fluid; an impermeable and hard "mud cake" or "filter cake" is thus formed on the wall of the borehole, when bentonitic drilling-mud is used.

Viscosity (or the resistance to flow) is an important property of the drilling fluid. In combination with the velocity of the circulated fluid, viscosity controls the ability of the fluid to remove cuttings from the borehole; but it has no relationship to density. In the field, it is measured by the time required for a known quantity of fluid to flow through an orifice of special dimensions. Most liquid drilling muds, either colloidal or emulsions that behave as plastic fluids (non-Newtonian), differ from those of (true) Newtonian fluids such as water and light oils, in that their viscosity is not constant, but varies with the rate of shear—on which in turn depends the ratio of shearing stress to the rate of shearing strain.

Yield point and **gel strength** are two additional properties that are considered in evaluating the characteristics of drilling mud. **Yield point** is a measure of the amount of pressure that must be exerted by the pump, upon restarting after a shutdown, to cause the drilling fluid to start to flow. **Gel strength** is a measure of the ability of the drilling fluid to maintain suspension of particulate matter in the mud column when the pump is shut down. There is a close relationship between viscosity, yield point, and gel strength (Assaad et al. 2003).

4.3.2 Composition and Nature of the Drilling Fluids

A typical drilling mud fluid consists of a continuous liquid base, a dispersed gel-forming phase (e.g., colloidal solids and/or emulsified liquids for the required viscosity), and wall cake; other inert dispersed solids (e.g., weighing materials, sand, and cuttings); and various chemicals necessary to control properties within desired limits. The general functions of drilling fluids are to cool and lubricate the bit and drill string; to remove and transport cuttings from the bottom of the hole to the surface; to suspend cuttings during times when circulation is stopped; to control encountered subsurface pressures; and to wall the borehole with an impermeable mud cake.

There are different kinds of drilling mud fluids:

(a) **Saltwater muds**—Sodium bentonite does not form a satisfactory colloid in salt water. The clay mineral attapulgite (known as salt clay), together with hydrates, forms a stable suspension in salt water and is used in saline water in about the same manner as bentonite in freshwater.

Salt concentrations neutralize the electric charge on dispersed bentonite particles, allowing the formation of particle aggregates which are larger than colloidal size (flocculation). Saltwater muds are used in drilling through salt beds and in localities of abundant saltwater (e.g., in offshore drilling, and in swamp and seaside locations). In general, the difference between freshwater and saltwater muds is the type of clay used in the gel-forming phase.

In offshore drilling operations, seawater, which is more economically used as the base fluid for the salt-mud system, is also effective in drilling shallow offshore domes.

(b) **Freshwater muds**—The basic ingredients are fresh water and suspended clays. Certain clay minerals, if ground to colloidal size and mixed with water, readily hydrate (adsorb water) to form stable colloids. Sodium bentonite, mainly composed of the clay mineral montmorillonite, yields higher viscosity at lower clay content than do other clay minerals.

It is a common practice to pretreat the mud system with calcium; such muds are known as lime base muds, in which calcium is allowed to remain in the mud because it tolerates other flocculating salts (up to 50,000 ppm of sodium chloride).

(c) **Oil base muds**—These are generally composed of varying materials: high flash diesel oil; oxidized asphalt (a colloidal fraction that provides the wall building property); a combination of an organic acid and an alkali (forming an unstable soap that governs the viscosity and gel strength of the mixture); and various stabilizing agents, plus 2–5% water.

Oil base muds are expensive, and their main uses are: (a) drilling and coring of possible productive zones; (b) drilling of bentonitic (heaving) shales that continually hydrate, swell, and slough into the hole when mixed with water; (c) deep drilling operations in high temperature environments where solidification may occur; (d) as a perforating fluid to be added (a few barrels) to prevent contamination of the section after being perforated; and (e) remedial operations on producing wells to avoid other drilling problems such as a stuck pipe and corrosion.

4.3.3 Mud Systems in Salt Structures

A complete technological cycle in mud engineering has been performed since Stroud (1925) introduced weighted drilling fluids to control blowouts. Mud systems became increasingly complex thereafter. Many organic and inorganic additives were added to gel muds to maintain the desired mud qualities under varying conditions. However, there has recently been a tendency to return to simpler mud systems that accomplish all the desired effects of former complex muds with fewer additives.

There are two basic types of mud systems—one designed for drilling shallow domes, and the other for deep salt structures; if salt is drilled at very shallow depths, a saturated salt water mud is mixed in the surface system to be utilized as a spud mud. However, if salt domes are several thousand feet deep (but still in the shallow or intermediate depth range), a freshwater mud system can be used until the cap salt rock is penetrated; then the mud is converted to a saturated salt mud. Both native or gel mud fresh water may require certain inorganic additives to develop the proper alkaline values which promote hydration and dispersion of the penetrated shales. The commonly used freshwater clays or gels are adversely affected by high salt concentration; therefore, a special saltwater clay, attapulgite, should then be used.

Saltwater muds have salt concentrations above 10,000 ppm, or one percent salt. Normal freshwater muds with either high or low pH values maintain good physical properties up to this concentration of salt. Saturated saltwater muds, which are characterized by high gel strength, are usually added to the system if abnormal viscosity and high gel strength create a problem (Halbouty 1967).

4.3.4 Salt Dome Drilling

Drilling on salt domes can be classified into three main types: (1) **salt dome drilling** on crests of shallow domes, where lost circulation is a common problem that occurs in drilling highly cavernous cap rock over most shallow domes; (2) **extended drilling capabilities to medium depths** on the flanks of shallow, piercement domes; and (3) **drilling of deep domal and anticlinal salt structures**, involving greater risks than shallow drilling and with many potential complications and problems—therefore, specialized mud systems must be programmed to ensure the best chance for a successful operation. Problems of deep drilling wells in or above salt structures include lost circulation, heaving shales, salt water flows, abnormally high pressure zones, and high temperature.

4.4 Drilling Hazards

Improper control of mud properties can create various drilling hazards:

(a) **Enlargement of boreholes in salt beds** of considerable thicknesses can cause very difficult fishing operations in case of drill string failure.

(b) **Heaving shale problems** in shale sections rich in bentonite or other hydrate clays continually adsorb water, swell, and slough into the hole; in other words, upon hydration, the shale forming the wall of the bore hole may disintegrate and fall into the hole, sticking the drill pipe and causing a severe drilling hazard. In hydratable shales, where heavy muds with a high content of suspended solids directly affect the viscosity, the control of viscosity by dilution requires the addition of large quantities of barite, which greatly raises the cost of a mud program.

(c) **Blowouts** occur when the encountered formation pressure exceeds the mud column pressure, allowing the formation fluids to blow out of the hole. Blowout preventers are connected to the top of the wellhead and are used to shut in and control the well in the event of a blowout of gas or oil, while drilling in the reservoir rock which encounters higher pressures than that exerted by the column of mud in the hole. Blowout preventers consist of two or three ram preventers and one big-type preventer; their working pressure is 3,000, 5,000, or 10,000 lb/sq inch, depending on the conditions of the well and the expected pressures of the reservoir.

(d) **Partial or complete lost circulation**, defined as the loss of substantial quantities of a part or of the whole mud, might occur in an encountered formation; though it is always required that the mud column pressure must exceed the formation pressure. Undesirable effects of lost circulation are expected, such as lack of lithological information, since no cuttings are obtained; the possibility of a stuck drill pipe, resulting in a fishing job; a consequent loss of drilling time; and cost increases.

4.5 Drill Stem Testing (DST)

4.5.1 Scope

A drill stem test is carried out when oil shows are traced in the cuttings; this can occur many times during the drilling of an exploratory well. It allows the productive capacity of an encountered formation to be tested in a borehole full of drilling mud. The testing tool assembly attached to the lower part of the string of the drill pipe (or drill stem) is lowered and placed opposite to the reservoir rock, isolated by an expandable packer(s), relieved of the mud column pressure, and allowed to produce through the drill stem. The hydrostatic pressure of the mud column inside the hole is always greater than the formation pressure of the production zone to be tested. The basic drill stem test tool assembly consists of the following (USEPS 1977, Edwards and Winn 1974, Kilpatrick 1955; Fig. 4.5 Black, Courtesy AIME):

1. A rubber packer(s) expanded against the hole to separate the annular sections above and below the encountered zone.

2. A tester valve (upper pressure recorder) to control mudflow into the drill pipe (or exclude mud) during

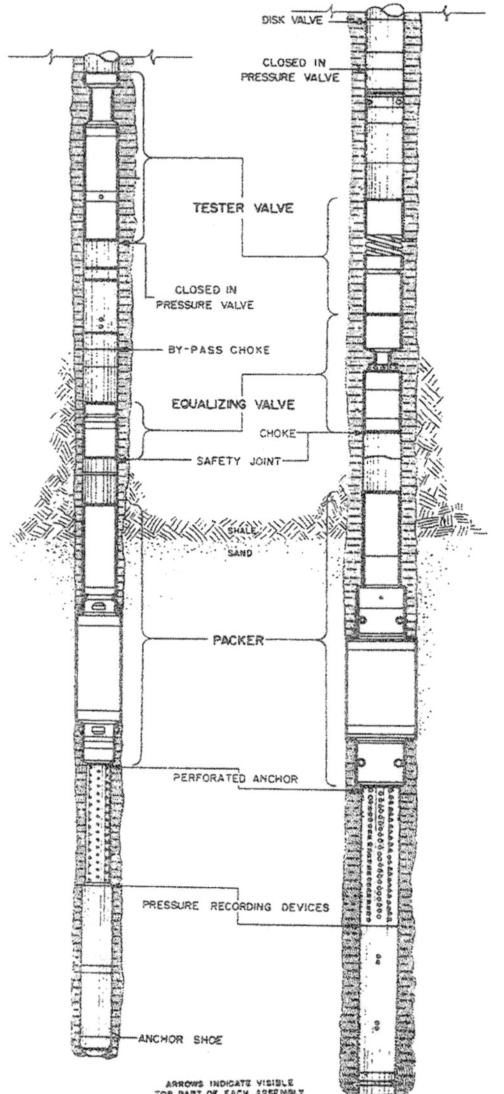

Fig. 4.5 Typical conventional drill-stem test tools (After Black, Courtsey AIME)

entry into the hole and to allow formation fluids to enter during the test. A hydraulic jar is fixed underneath the tester valve to increase the pressure to 8 psi, for releasing the packer after the test.

3. An equalizing bypass valve to allow mud pressure equalization across the packer(s) in and out of the DST assembly after completion of the flow test.

4. A pressure recorder for obtaining quantitative results during the period of flow in pressure (FIP) or pressure building (close-in pressure, CIP) following the period in which the formation fluids are allowed

(FIP). Pressure recorders furnish a complete record of all events that may occur during a particular test and are in the form of a graph of pressure versus time.

Two pressure recorders are usually located to measure the pressure inside and outside the perforated anchor. The two measurements allow accurate determination of whether or not the perforations have become plugged during the test.

A perforated anchor is the extension below the tool that supports the weight applied to set the packer. It rests either on the bottom of the hole in open-hole tests or on cement plugs that have been spotted at the desired location. Safety joints are equipped to afford a means of unscrewing the drill string at a point convenient for fishing operations, in case the packers become stuck. Another safety joint is added underneath to prevent the DST assembly from sticking.

4.5.2 Discussion

The Halliburton Formation Testing Procedure was the first known method to determine the potential productivity of a reservoir rock in either an open or a cased hole. The testing procedure requires the opening of a section of the borehole to atmospheric or reduced pressure. The testing string is lowered into the hole on a drill pipe with the tester valve closed to prevent entry of well fluid into the drill pipe, leading to an undesired fishing job and possibly a stuck drill pipe.

The procedure for testing the bottom section of a borehole can be summarized as follows (Halliburton; Fig. 4.6a, modified):

(a) While going in the hole, the packer is collapsed, allowing the displaced mud to rise, as shown by the arrows (Fig. 4.6a, step #1).

(b) After the drill stem reaches bottom, and the necessary preparations are completed, the packer is then set (compressed and expanded) and isolates the lower zone (or the desired zone) from the rest of the open hole; in other words, it provides a seal above the zone to be tested (step #2).

(c) The bypass is closed as the tester valve is opened; here, the packer supports the hydrostatic pressure load of the well fluid, and the isolated section is exposed through the open tester valve to the low

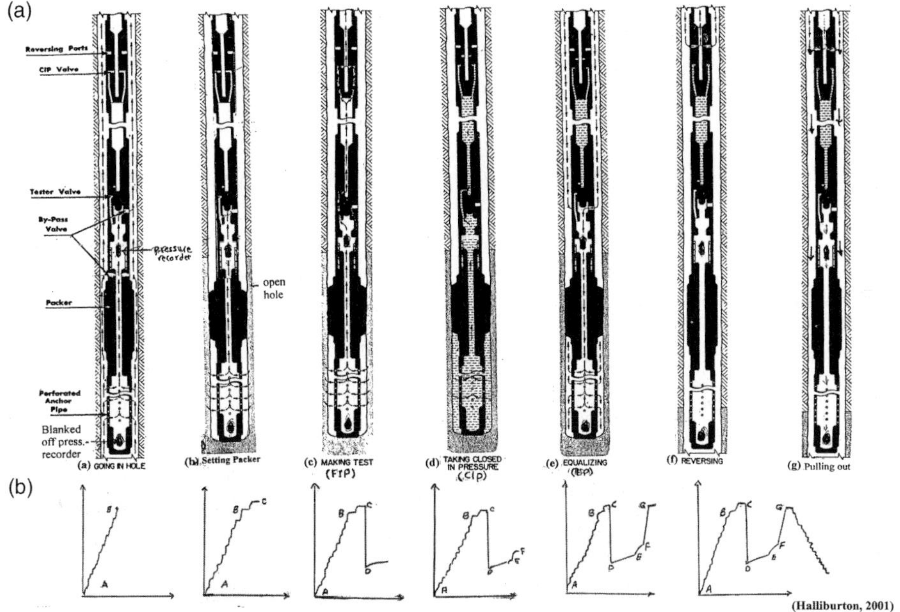

Fig. 4.6 (**a**) A detailed illustration of drill stem testing. (**b**) Fluid passage diagram for a conventional bottom section, drill stem test. Courtesy Halliburton Oil Well Cementing Company (modified by Assaad)

pressure inside the empty or nearly empty drill pipe, allowing the formation fluid to enter and the flowing formation pressure (FIP) can be measured during the flow period (step #3).

At the end of the test, the tester valve is closed (CIPV), trapping any fluid above it; this makes possible the measurement of the static formation's "built-up in-pressure (CIP)" (step #4).

d) After the final closed-in period, the bypass valve is opened to equalize the pressure across the packer (EP) (step #5).

e) Formation fluid received during the test can be removed from the drill pipe by reverse circulation before the pipe is removed from the borehole. This reversal is performed by closing the blowout preventers and pumping mud down the annulus; the mud then enters the drill pipe through the reversing ports, thereby displacing any formation fluids in the pipe. The recovered fluids may be sampled as they are discharged at the surface (step #6).

Finally, the setting weight is taken off and the packer is pulled free (step #7). The fluid content of each successive pipe section is examined when it is removed. The graphic charts are absolutely essential to get the accurate interpretation of test results. Fig. 4.6b shows a record of fluid passage, a graph of pressure versus time for a conventional bottom section of a drill stem test (Halliburton 2001). Tests through perforations are sometimes required to retest a reservoir rock after casing has been emplaced. Fig. 4.7, 4.8, 4.9, 4.10, 4.11, 4.12, 4.13 and 4.14, show common types of drill stem testing.

4.6 General Remarks

4.6.1 About the DST—Procedure

The following is a DST procedure to further explain the steps given in Fig. 4.6a:

(1) Initial hydrostatic pressure (IHP) is exerted by mud column. (2) Initial closed-in pressure (ICIP). (3) Initial flow pressure (IFP) is the lowest pressure recorded just after the tool is opened. (4) Final flowing pressure (FFP) is the pressure just before the tool is closed. (5) Final closed-in pressure (FCIP). (6) Final hydrostatic in-pressure (FHP).

– The choke size of the bottom hole surface orifices selected depends on the test's anticipated

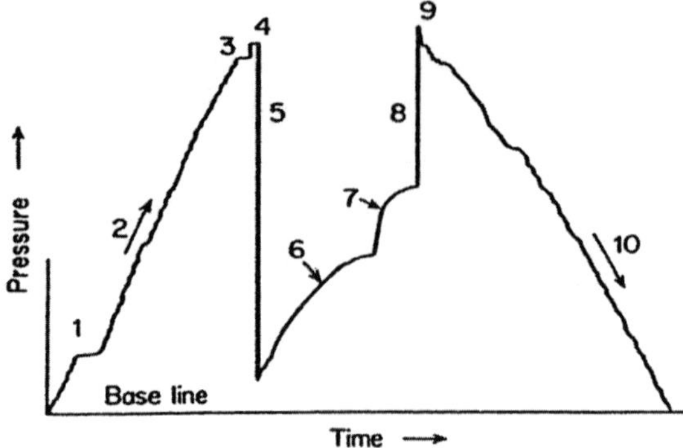

Fig. 4.7 Normal sequence of events in successful drill stem test. After Kirkpatrick, courtesy *Petroleum Engineer*

1. Putting water cushion in drill pipe
2. Running in hole
3. Hydrostatic pressure (weight of mud column)
4. Squeeze created by setting packer
5. Opened tester, releasing pressure below packer
6. Flow period, test zone producing into drill pipe
7. Shut in pressure, tester closed immediately above packer
8. Equalizing hydrostatic pressure below packer
9. Released packer
10. Pulling out of hole

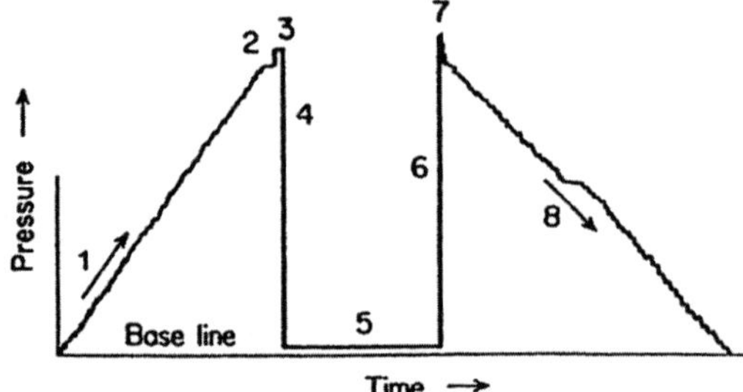

Fig. 4.8 *Dry* test. After Kirkpatrick,[1] courtesy *Petroleum Engineer*

1. Running in hole
2. Hydrostatic pressure (weight of mud column)
3. Squeeze created by setting packer
4. Opened tester, releasing pressure below packer
5. Flow period, test zone open to atmosphere
6. Closed tester and equalizing hyd. pressure below packer
7. Pulled packer loose
8. Pulling out of hole

conditions; the bottom choke is used primarily as a safety measure and should be considerably large enough to minimize surface pressure in a flowing test conditions.

– The water cushion of a certain length or head of liquid is placed inside the drill pipe rather than running it dry, to reduce the external pressure on the drill pipe in deep holes and/or to reduce the pressure drop on the formation and across the packer(s) when the DST assembly is first opened.

– The length of the DST depends greatly on some observations during the test; e.g., in hard formations, the drill pipe is not liable to be stuck and the flow test is often several hours long; the shut-in period after the flow test should be long enough to permit establishment of a stabilized static pressure.

4.6.2 Required Conditions and Reasons for Carrying Out a DST in a Petroleum Reservoir Formation

• The formation pressure of oil or gas should be maintained during drilling and the mud fluid pressure should exceed that of the formation.
• Saturated cuttings and core samples of gas or oil.
• Porous and permeable core samples.
• Presence of oil or gas in the drilling mud.

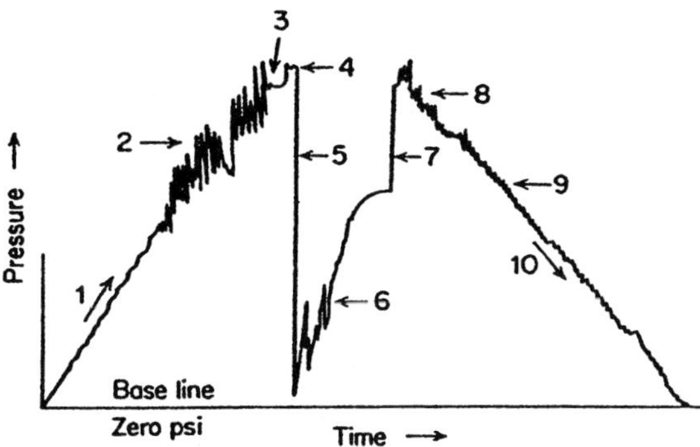

Fig. 4.9 Effect of poor hole condition. After Kirkpatrick, courtesy *Petroleum Engineer*

1. Running in hole
2. Indicates bad hole condition, scraping wall cake
3. Hydrostatic pressure (weight of mud column)
4. Squeeze created by setting packer
5. Opened tester, releasing pressure below packer
6. Indicates plugging of perf. anchor or tester
7. Closed tester and equalizing hyd. pressure below packer
8. Indicates swabbing due to bad hole condition
9. Indicates less swabbing, better hole condition
10. Pulling out of hole

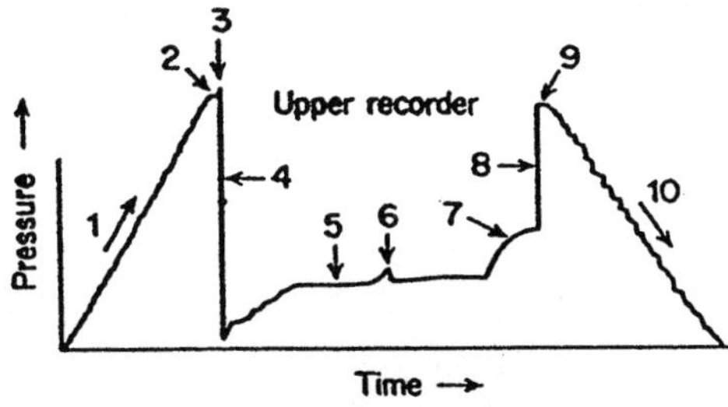

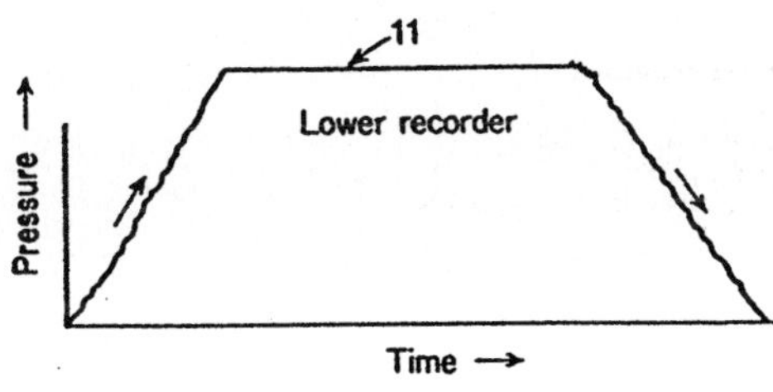

1. Running in hole
2. Hydrostatic pressure (weight of mud column)
3. Set packer
4. Opened tester, releasing pressure between packers
5. Test zone flowing
6. Rise in pressure due to closing in to change chokes
7. Shut in bottom hole pressure
8. Equalizing hydrostatic pressure between packers
9. Hyd. pressure at conclusion of test
10. Pulling out of hole
11. Lower recorder, below bottom packer shows no drop in pressure, proves bottom packer is holding

Fig. 4.10 Straddle packer test. After Kirkpatrick, courtesy *Petroleum Engineer*

- Shows of fluorescence in cuttings and core samples.
- Identification of petroleum from electric log interpretations.

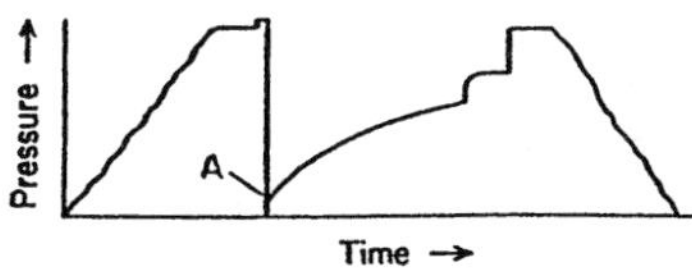

It will be noted from point A on the chart above that the pressure line indicating the opening of the tool, coincides with the start of the build-up in flow pressure. The exact initial flow pressure point cannot be determined. This condition is caused by the successive pressure events happening too rapidly to allow the clock to turn the chart and separate the lines.

Fig. 4.11 Effect of pressure recorder inertia. After Kirkpatrick, courtesy *Petroleum Engineer*

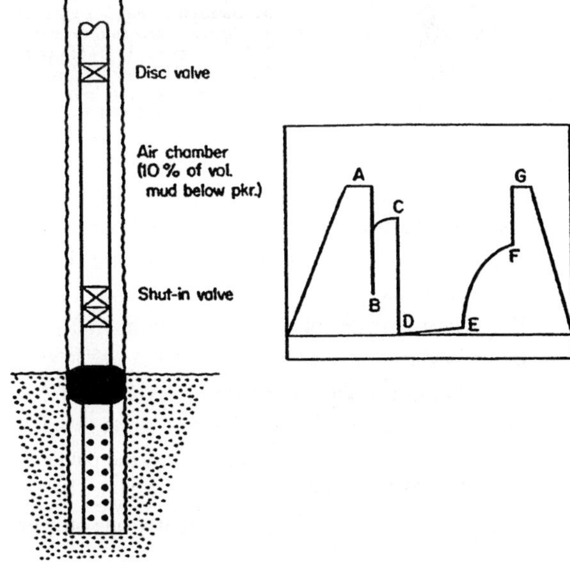

Fig. 4.12 Schematic tool arrangement for procuring initial closed-in pressure. Point C on chart is taken as reservoir pressure. Courtesy Johnson Testers

Fig. 4.13 Double Pressure charts – Anchor plugged

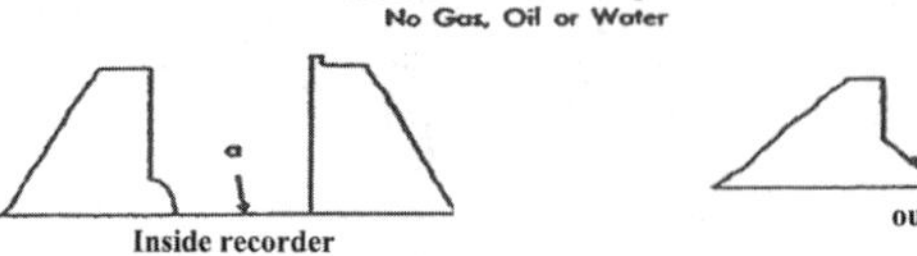

Fig. 4.14 Double Pressure charts – Choke plugged

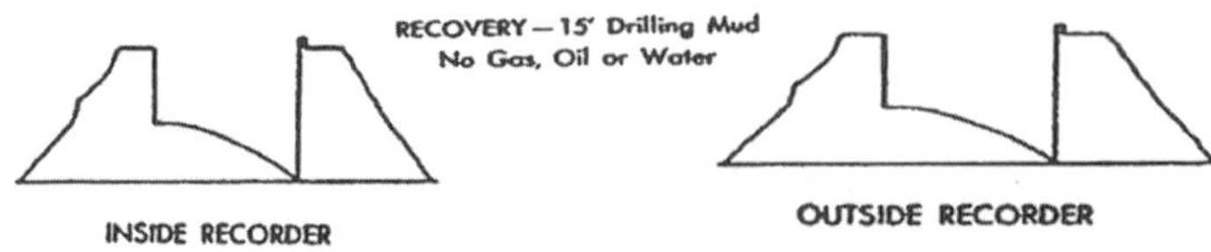

Appendix 4.A

Selected Graphs of Drill Stem Testing in Different Reservoir Rocks and Interpretations (Figs. 4.15, 4.16, 4.17, 4.18, 4.19, 4.20, 4.21, 4.22, 4.23, 4.24, 4.25, 4.26, 4.27, 4.28, 4.29, 4.30, 4.31, 4.32, 4.33, 4.34, 4.35, 4.36, 4.37, 4.38, 4.39.

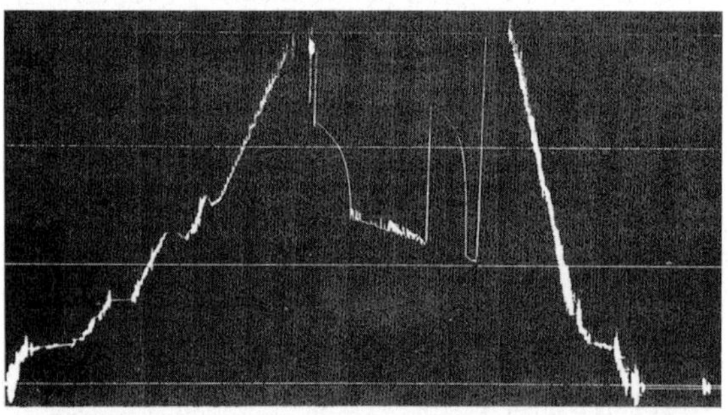

Fig. 4.15 Incorrect base line and shut off during flow (Ligne de base incorrecte et bouchage pendant débit)

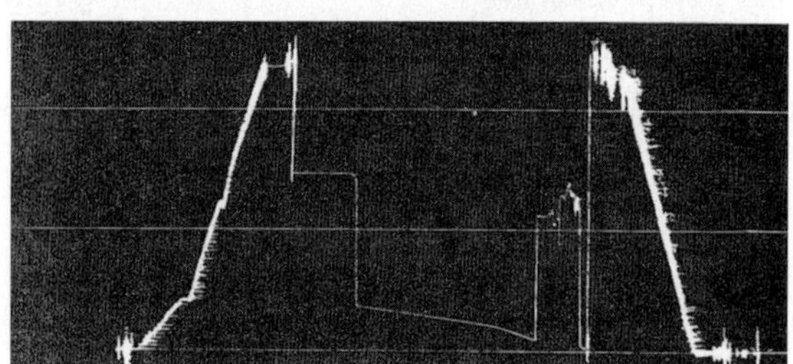

Fig. 4.16 Leaking dual closed—in-pressure valve "DCIPV" (Fuite au Vanne de fermeture et ouverture)

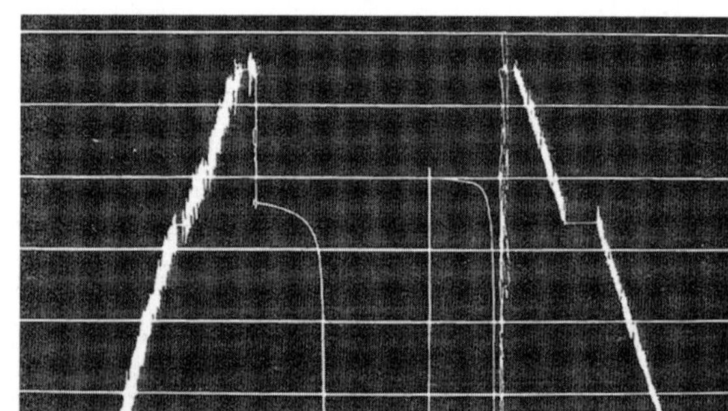

Fig. 4.17 Clock running away (Montre marchant très vite)

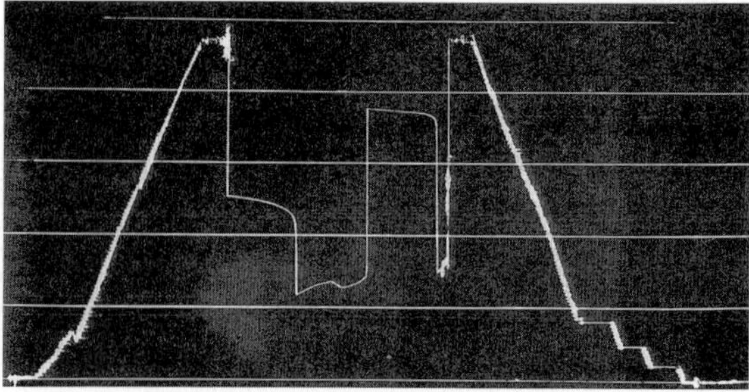

Fig. 4.18 Initial rapid built up pressure (Premiere remontèe de pression surcharge)

Fig. 4.19 Depletion of gas reservoir (Reservoir de gaz en depletion)

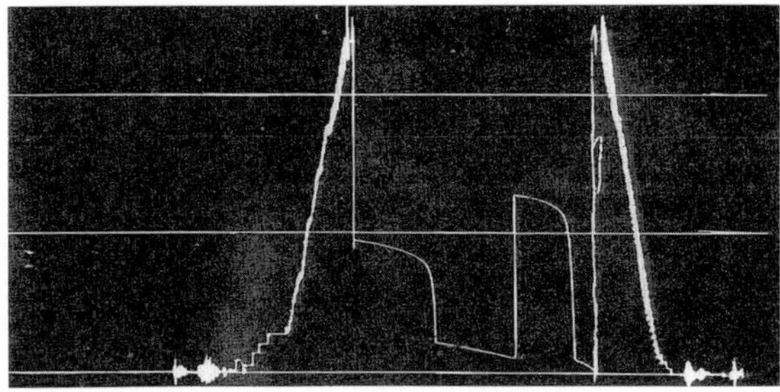

Fig. 4.20 Depletion of oil reservoir (Reservoir de liquide en depletion)

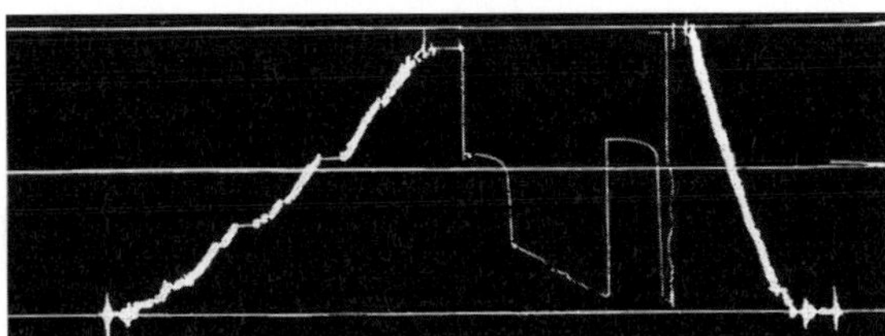

Fig. 4.21 Plugging flow period with a uniform segment (Bouchage pendant dèbit)

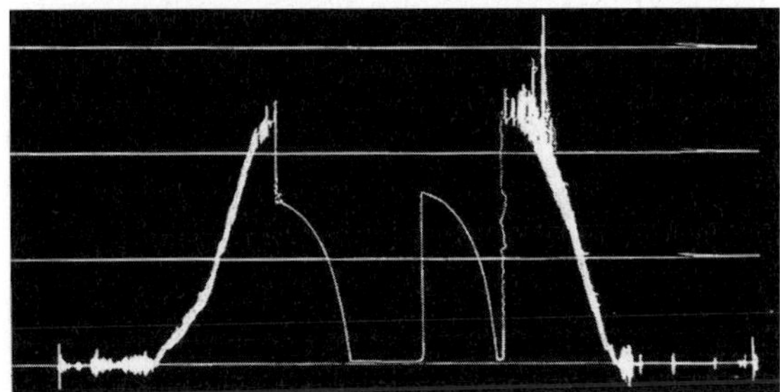

Fig. 4.22 Low permeability formation with high reservoir pressure (Formation á faible permèabilitè avec grande pression de reservoir)

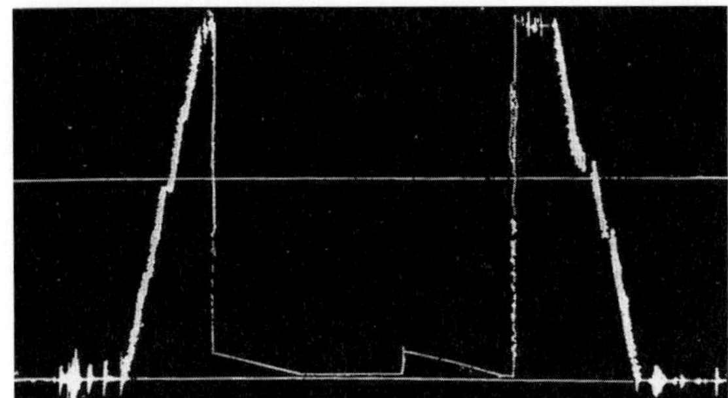

Fig. 4.23 Low permeability formation with low reservoir pressure (Formation á faible permèabilitè avec grande faible pression de reservoir)

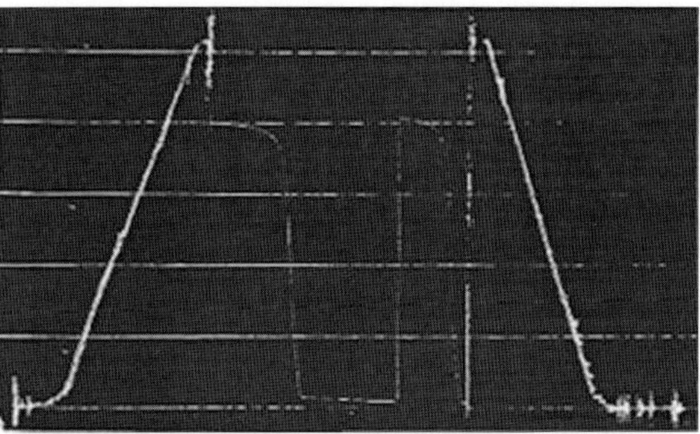

Fig. 4.24 The "S" curve indicates a possible after-production effect of bypass gas (Courbe en "S" indiquant l'effet d'une venue de gaz possible apres-production)

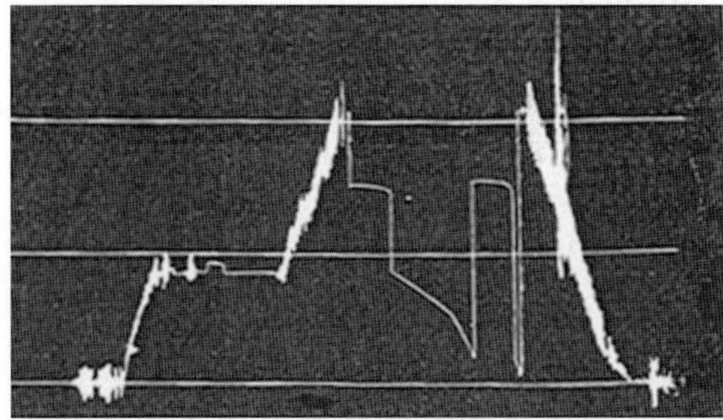

Fig. 4.25 Transition of filling up of the drilling pipes of the well (Transition de remplissage de massestiges aux tiges de forage)

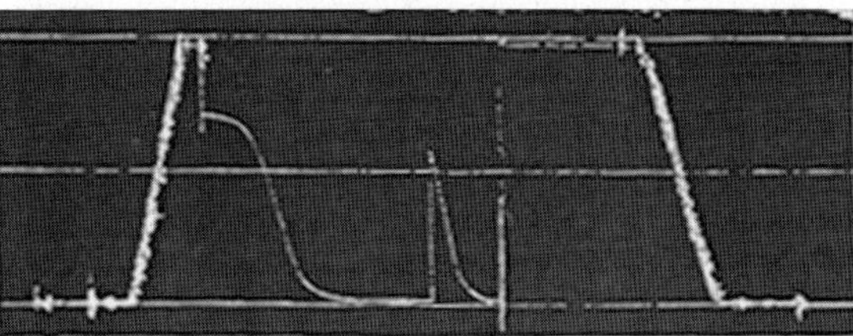

Fig. 4.26 The "S" curve indicates a possible vertical permeability (Courbe en "S" indiquant une permèabilitè verticale possible)

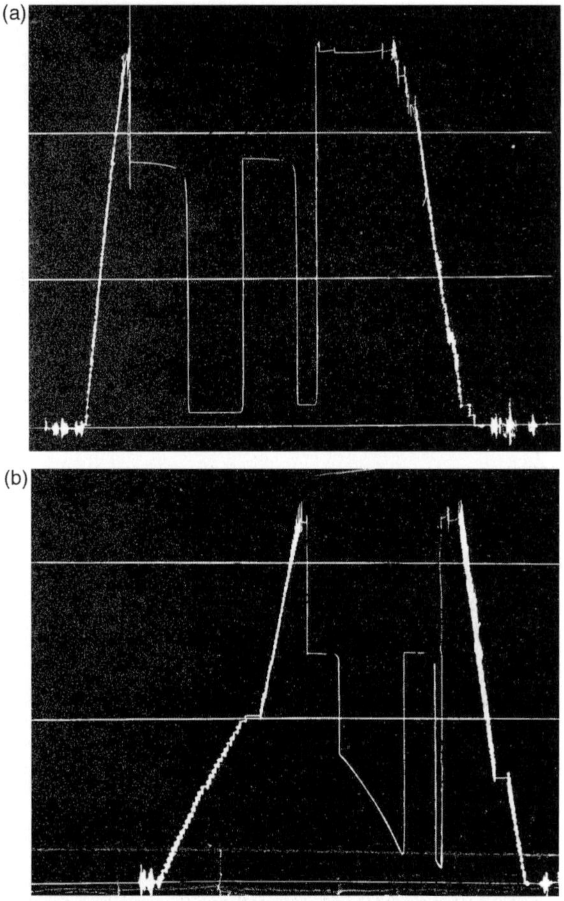

(a)

(b)

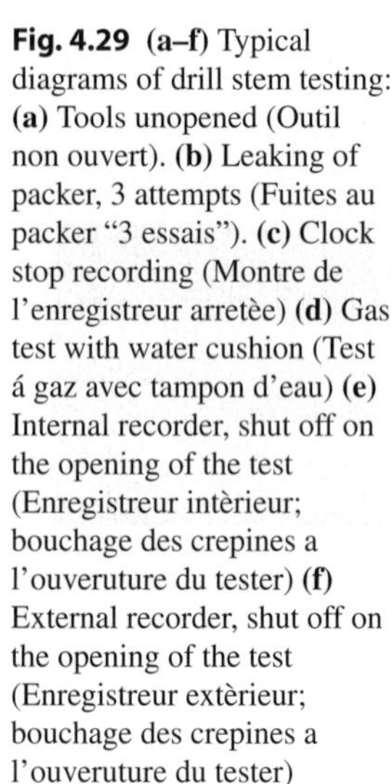

Fig. 4.27 **(a)** Low productivity and high damage (Faible productivitè + grande endommagement) **(b)** A high productivity and high damage (Grande productivitè + Grande endommagement)

(a)

(b)

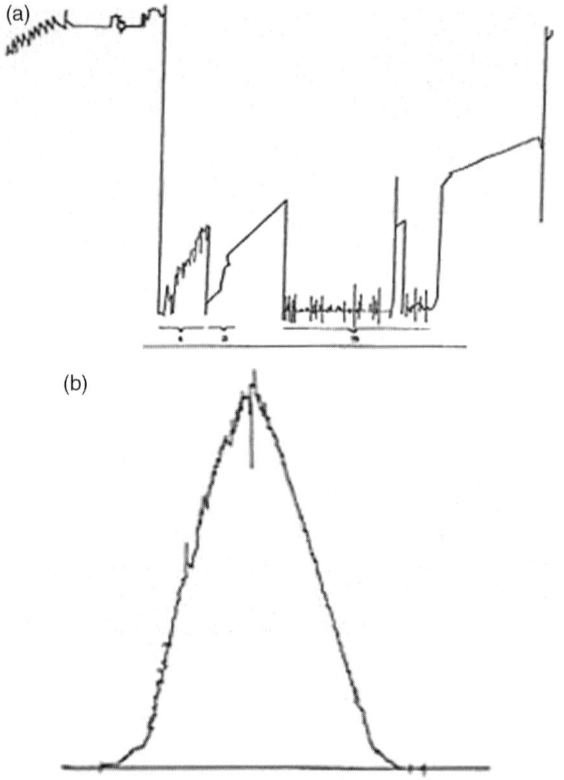

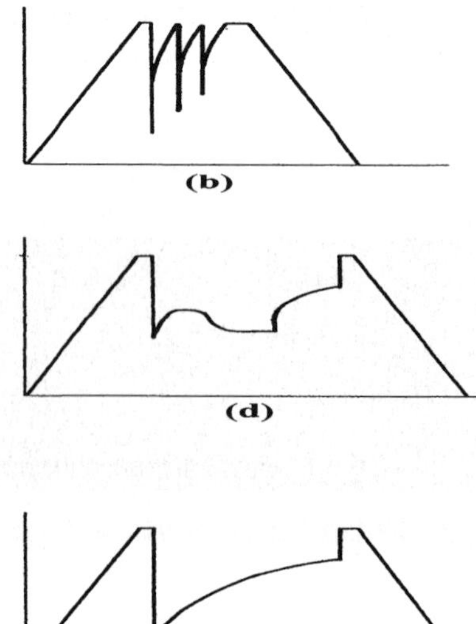

Fig. 4.28 **(a)** 1—Sliding of packer; 2—reanchoring of packer/test opened/and sliding of packer; 3—partial plugging of "fringes" (1— Glissement du packer; 2—Reancrage du packer/Test ouvert/et glissement de packer; 3—Bouchage partial des crepines). **(b)** Leaking of packer (Fuite au packer)

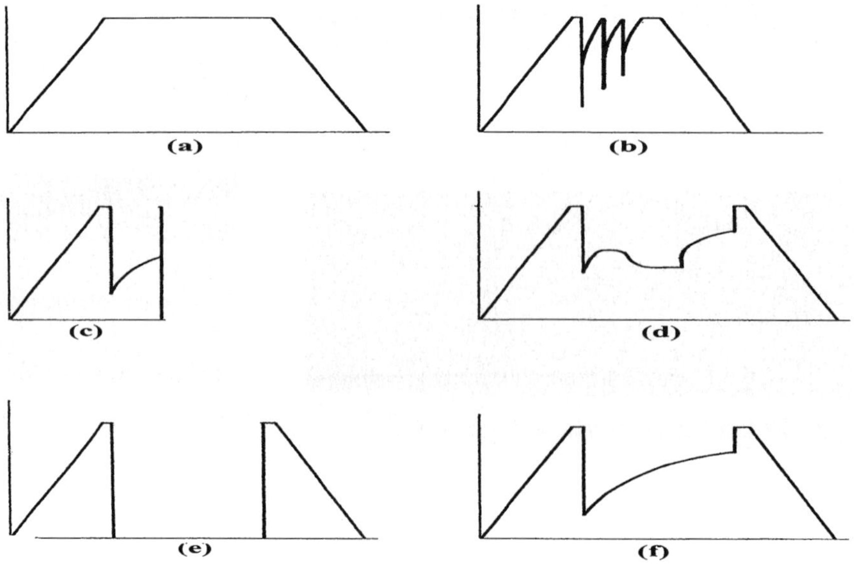

Fig. 4.29 **(a–f)** Typical diagrams of drill stem testing: **(a)** Tools unopened (Outil non ouvert). **(b)** Leaking of packer, 3 attempts (Fuites au packer "3 essais"). **(c)** Clock stop recording (Montre de l'enregistreur arretèe) **(d)** Gas test with water cushion (Test á gaz avec tampon d'eau) **(e)** Internal recorder, shut off on the opening of the test (Enregistreur intèrieur; bouchage des crepines a l'ouveruture du tester) **(f)** External recorder, shut off on the opening of the test (Enregistreur extèrieur; bouchage des crepines a l'ouveruture du tester)

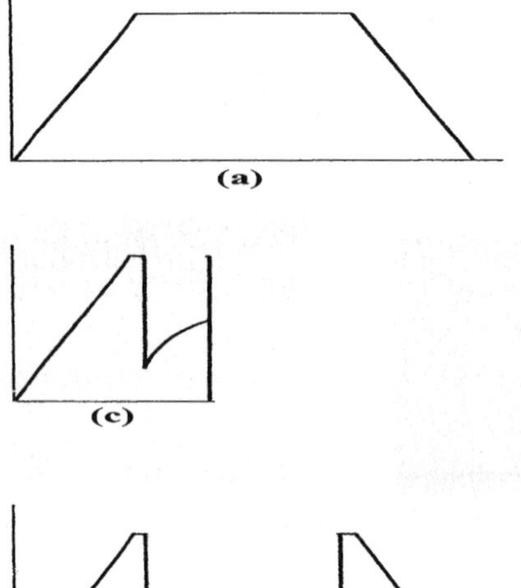

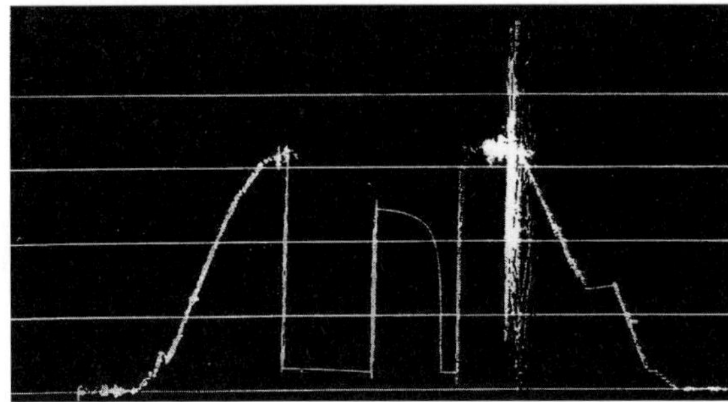

Fig. 4.30 Leaking of the drilling pipes. (Trou ètroit indiquè et fuite au tiges de forage)

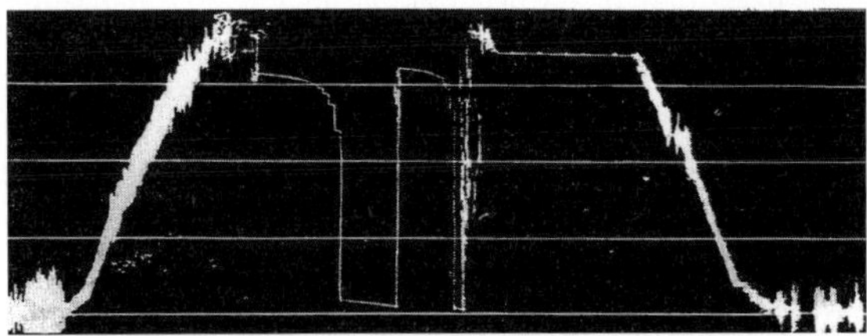

Fig. 4.31 A recorder of stair stepping (Enregistreur: stair stepping)

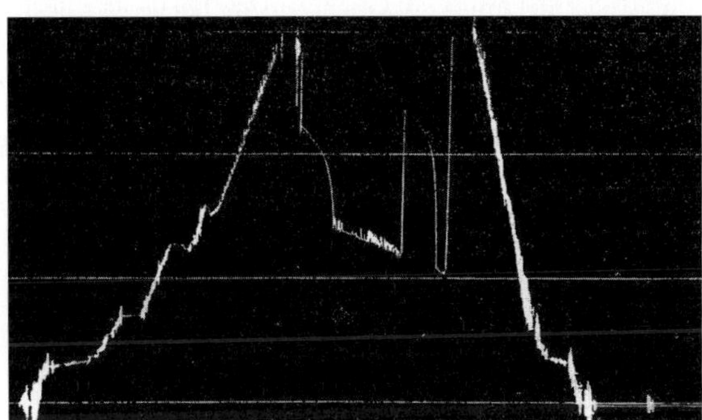

Fig. 4.32 Incorrect base line and plugging during flow (Ligne de base incorrecte et bouchage pendant dèbit)

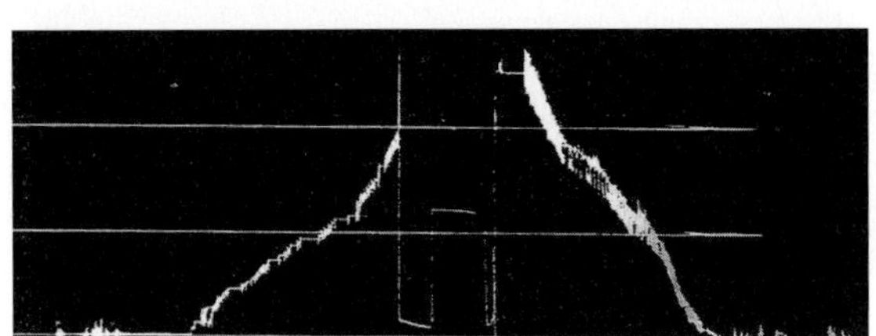

Fig. 4.33 Clock stopped (Arrêt de la montre)

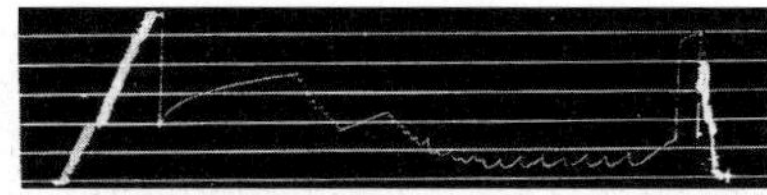

Fig. 4.34 Swapping (Pistonage)

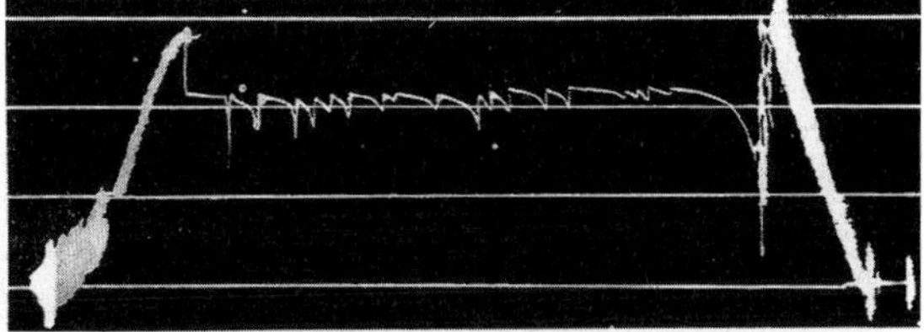

Fig. 4.35 Plugged formation through flow testing (Débit bouché)

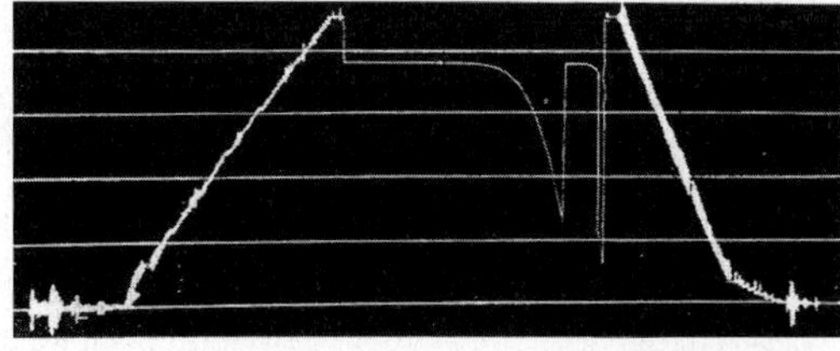

Fig. 4.36 Equalized flow (FSIP = FFP) (Débit égalisé)

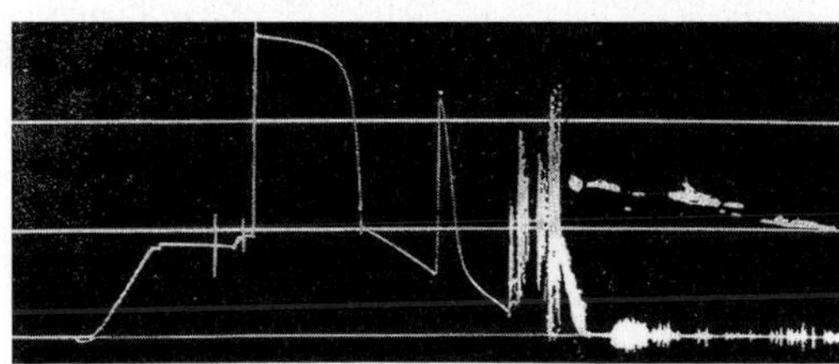

Fig. 4.37 The after flow pressure conceals ISIP (Courbe en "S" dévellope par compression de fluide due a l'ancrage du packer assez plus haut)

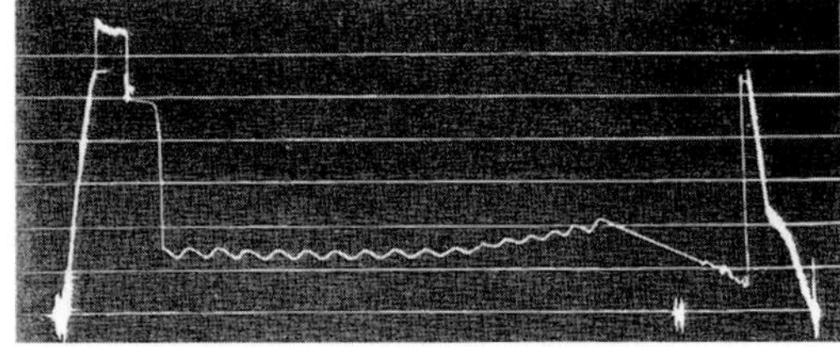

Fig. 4.38 Flowing in leads (Dèbit en coup)

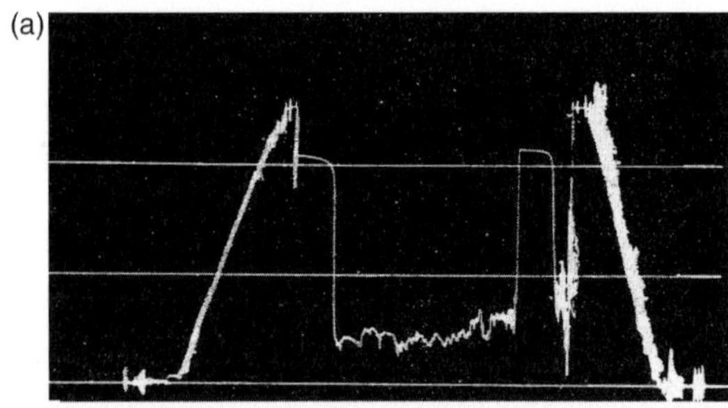

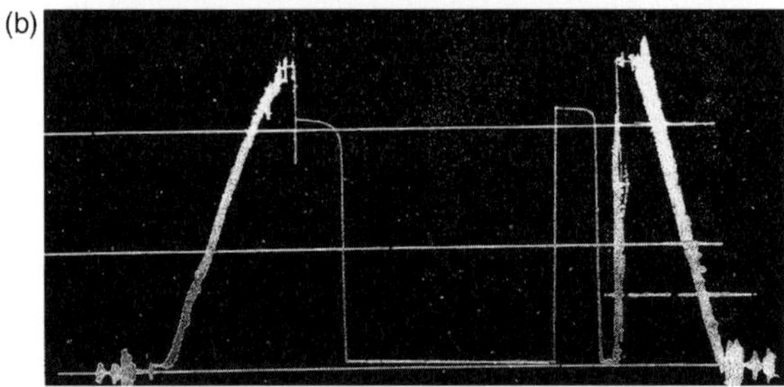

Fig. 4.39 Bottom pressure measurement show plugging of perforations (Diagramme enregistreur du bottom). **(b)** Top pressure measurement show plugging of perforations (Diagramme enregistreur du top)

Appendix A

Figs. 4.15, 4.16, 4.17, 4.18, 4.19, 4.20, 4.21, 4.22, 4.23, 4.24, 4.25, 4.26, 4.27, 4.28, 4.29, 4.30, 4.31, 4.32, 4.33, 4.34, 4.35, 4.36, 4.37, 4.38, 4.39.

Selected Graphs Showing Fluid Pressures during Drill Stem Testing Procedure in Different Reservoir Rocks

References

Assaad F et al (2003) Field Methods for Geologists and Hydrogeologists, Drilling and Testing–Soil samplers, Ch(6)/p122 Springer, Heidelberg, Germany

Edwards AG and Winn RH (1974) A summary of modern tools and techniques used in drill stem testing; presented at the dedication of the US East-West trade center. Tulsa, ORlahoma, EPA

Kilpatrick CV (1955) An integrated summary of formation evaluation criteria, AIME T.P. 595-G, presented at formation evaluation symposium, Houston Univ.

Brantly JE (1952) Rotary drilling handbook, 5th ed. New York, Palmer Publications, pp10–235 *in* Petroleum Engineering, Gatlin (1960); Fig 4.1/p41

Eastman Oil Well Survey Co., *in* Petroleum Engineering, Gatlin (1960), Fig 9.21/p157

API, R.P. 9B, 2nd (1954) American Petroleum Institute, New York, *in* Petroleum Engineering, Gatlin (1960); Fig 5.1/p53

Main WC (1949) Detection of incipient drill-pipe failures. API drilling and production recommended practices, No. 29

Black WM (1956) A review of drill-stem testing techniques and analysis. Jr. of Petroleum Technology, p21, *in* Petroleum Engineering, p256/Fig 13.3, Gatlin, (1960)

Kirkpatrick CV (1954) Formation testing—the petroleum engineer, pB-139; Figs 13.5–13.9/p258–259 & Fig 13.11/p260: *in* Petroleum Engineering, Gatlin (1960)

Edwards SH and Miller CP (1939) Discusssion on the effect of combined longitudinal loading and external pressure on the strength of oil well casing. API Drilling and Production Practices, p483

Halbouty MT (1967) Salt domes—Gulf Region, United States and Mexico. Figs. 9.1/p145; Gulf Publishing Co., P.O. Box 2608, Houston, Texas 77001

Stroud BK (1925) Use of barytes as a mud-laden fluid. *In* Oil World, p29

Chapter 5

Geophysical Well Logging Methods of Oil and Gas Reservoirs

5.1 Introduction

Several texts, manuals, field books, and guidebooks have discussed different geophysical log methods to define productive and nonproductive reservoirs in the petroleum business.

The properties of penetrated formations and their fluid contents are recorded by geophysical logs to measure their electrical resistivity and conductivity, their ability to transmit and reflect sonic energy, their natural radioactivity, their hydrogen ion content, their temperature and density, etc. The logs are then interpreted in terms of lithology, porosity, and fluid content (Table 5.1, USEPA 1977).

The basic concepts of well log interpretations and the factors affecting logging measurements are presented by Asquith and Gibson (1982) in several examples of oil and gas reservoirs, together with log interpretations given by Schlumberger (1972) and Dresser Industries (1975).

5.2 Borehole Parameters and Rock Properties

The parameters of log interpretations are determined by electrical, nuclear, or sonic (acoustic) logs. Rock properties that affect logging measurements are porosity, permeability, water saturation, and resistivity, and these are used in terms and symbols in well log interpretations (Schlumberger 1974).

Table 5.1 Geophysical well logging methods and practical applications (USEPA 1977)

	Method	Property	Application
1	Spontaneous Potential(SP) log	Electrochemical and electrokinetic potentials	Formation water resistivity; (Rw) shaliness (sand or shale).
	Non-focused electric log	Resistivity	Water and gas/oil saturation; porosity of water zones; Rw in zones of known porosity; formation resistivity (Rt); resistivity of invaded zone (Ri).
3	Focused and non-focused micro- resistivity logs	Resistivity	Resistivity of the flushed zone (Rxo); porosity; bed thickness.
4	Sonic log	Travel time of sound	Rock permeability.
5	Caliper log	Diameter of borehole	Without casing.
6	Gamma Ray	Natural radioactivity	Lithology (shales and sands).
7	Gamma-Gamma	Bulk density	Porosity, lithology.
8	Neutron-Gamma	Hydrogen content	Porosity with the aid of hydrogen content

F.A. Assaad, *Field Methods for Petroleum Geologists*,
© Springer-Verlag Berlin Heidelberg 2009

1. **Borehole diameter (d_h)** is the outside diameter of the drill bit and describes the borehole size of the well.
2. **Drilling Mud (R_m)** is a circulating fluid having a special viscosity and density to help remove cuttings from the well bore, lubricate and cool the drill bit, and keep the hydrostatic pressure in the mud column greater than the formation pressure.
3. **Mud filtrate (R_{mf})** is the fluid of the drilling mud that filters into the formation during invasion, whereas its solid particles of clay minerals are trapped on the side of the borehole and form the mud cake (R_{mc}).
4. **Water saturation** represents an important log interpretation concept because it can determine the hydrocarbon saturation of a reservoir by substracting the water saturation from the value of 1 (where $1.0 = 100\%$ water saturation):

$$\text{Water Saturation (Sw)} = \frac{\text{formation of water occupying pores}}{\text{Total pore space in the rock}}$$

5. **Invaded Zone**—consists of a flushed zone (R_{xo}) "close to the borehole, a few inches from the well bore" where the mud filtrate has completely flushed out the formation's hydrocarbons; **a transition zone or annulus zone** (R_j), where a formation's fluids and mud filtrate are mixed, occurs between the **flushed zone** (R_{xo}) and the **uninvaded zone** (R_t). The depth of invasion of the mud filtrate into the invaded zone is referred to as the diameter of invasion "di and dj" and depends on the permeability of the mud cake and not upon the porosity of the rock (Fig. 5.1; Schlumberger 1977, AAPG 1982).

NB: An equal volume of mud filtrate can invade low porosity and high porosity rocks, if the drilling muds have equal amounts of solid particles which coalesce and form an impermeable mudcake, acting as a barrier to further invasion.

When oil is present in the flushed zone, the degree of flushing by mud filtrate can be determined from the difference between water saturations in the flushed (S_{xo}) zone and the uninvaded (S_w) zone. Usually, about 90% of the oil is flushed out ($S_{xo} = 90\%$); the remaining oil is called residual oil saturation (ROS):

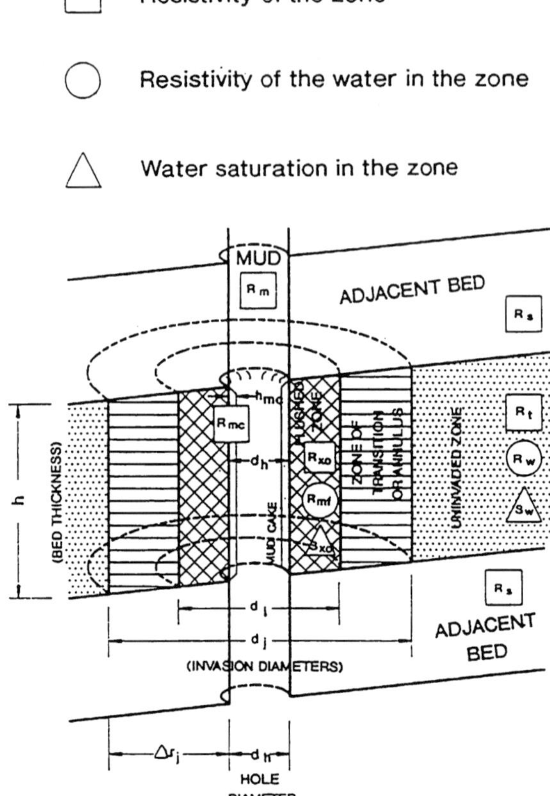

Fig. 5.1 A Diagram Shows the Invasion of Fluids through the Surrounding rock: The Cylindrical nature of the Invasion is also Shown by Dashed lines. After Asquith, G. and Gibson, C., 1982. "Courtesy Schlumberger Wall Services, Copyright 1971." (AAPG, 1982)

Degree of flushing by mud filtrate (ROS);

$$S_{10} = \{S_w - S_{xo}\}; \text{ or } S_{10} = (1.0 - S_{xo}).$$

6. **Uninvaded zone (R_t)**—is located beyond the invaded zone. In hydrocarbon-bearing reservoirs, there is always a layer of formation water on grain surfaces. Water saturation of the uninvaded zone (S_w) is an important factor in hydrocarbon reservoir evaluation:

$$S_h = \{1.0 - S_w\}.$$

Where:

S_h = hydrocarbon saturation, or the fraction of pore volume filled with hydrocarbons.

S_w = water saturation in the uninvaded zone, or the fraction of pore volume filled with formation water.

The ratio between the water saturation of the uninvaded zone (S_w) and that of the flushed zone (S_{xo}) is an index of hydrocarbon moveability.

5.3 Resistivity Measurements by Well Electric Logs

5.3.1 Definition

Resistivity, measured by electric logs, is a basic measurement of a reservoir's fluid saturation and is a function of porosity, type of fluid, (e.g., hydrocarbons, salt, or fresh water), and rock type; resistivity measurements are used to detect hydrocarbons and estimate the porosity of the reservoir. Both rock and hydrocarbons act as insulators, but salt water is conductive. Resistivity is measured in ohm-meters:

$$\mathbf{Resistivity\ R} = \frac{r.A}{L} \qquad (5.1)$$

where:

R = resistivity (ohm-meter$_2$/meter- or ohm- meters)
r = resistance (ohm)
A = cross sectional area of measured substance (m^2)
L = length of substance being measured (ms).

Resistivity can be used by logging tools to detect hydrocarbons and estimate the porosity of the reservoir. Cornard Schlumberger in 1912 began the first experiments which led eventually to the development of the modern day petrophysical logs; the first electric log was run by a French engineer in 1927, and in 1942 Archie with Shell Oil Co. presented a paper to the AIME in Dallas, Texas, which set forth the concepts used as the basis for modern quantitative log interpretation.

Archie submitted an experiment that the resistivity of a water-filled formation (R_0), filled with water of a resistivity R_w, can be related by means of a formation resistivity factor:

$$\mathbf{Archie's\ formula} : R_0 = F \times R_w \qquad (5.2)$$
$$\text{or, } F = R_0/R_w$$

where:

R_0 = resistivity of water-filled formation (100% water saturated);
F = Formation resistivity factor;
R_w = Formation water resistivity.

Also, Archie's formula stated that formation factors can be related to porosity by:

$$F = 1.0/\emptyset^m (m = \text{cementation exponent}). \qquad (5.3)$$

The cementation exponent varies with grain size and its distribution, and with the complexity of the paths between pores, known as turtuosity, for which the value is proportional with the m value.

5.3.2 Annulus and Resistivity Profiles—Hydrocarbon Zone

Theannulus profile, which reflects a temporary fluid distribution, occurs between the invaded zone and the uninvaded zone, denotes the presence of hydrocarbons, and can be detected only by an induction log run soon after a well is drilled; however, it should disappear with time. In the annulus zone, beyond the outer boundary of the invaded zone, the pores are filled with residual hydrocarbons (R_h) and formation water (R_w), and results in an annulus profile that causes an abrupt drop in measured resistivity. The annulus profile is important for petroleum geologists because it only appears in hydrocarbon bearing zones. In such cases, the mud filtrate invades hydrocarbons, which move out first; then formation water is pushed out in front of the mud filtrate, forming an annular (circular) ring at the edge of the invaded zone. Log resistivity profiles illustrate the values of the invaded and uninvaded zones.

Figure 5.2 shows a horizontal section through a permeable hydrocarbon-bearing formation and the related resistivity profiles which occur when there is invasion by either freshwater- or saltwater-base muds (AAPG 1982):

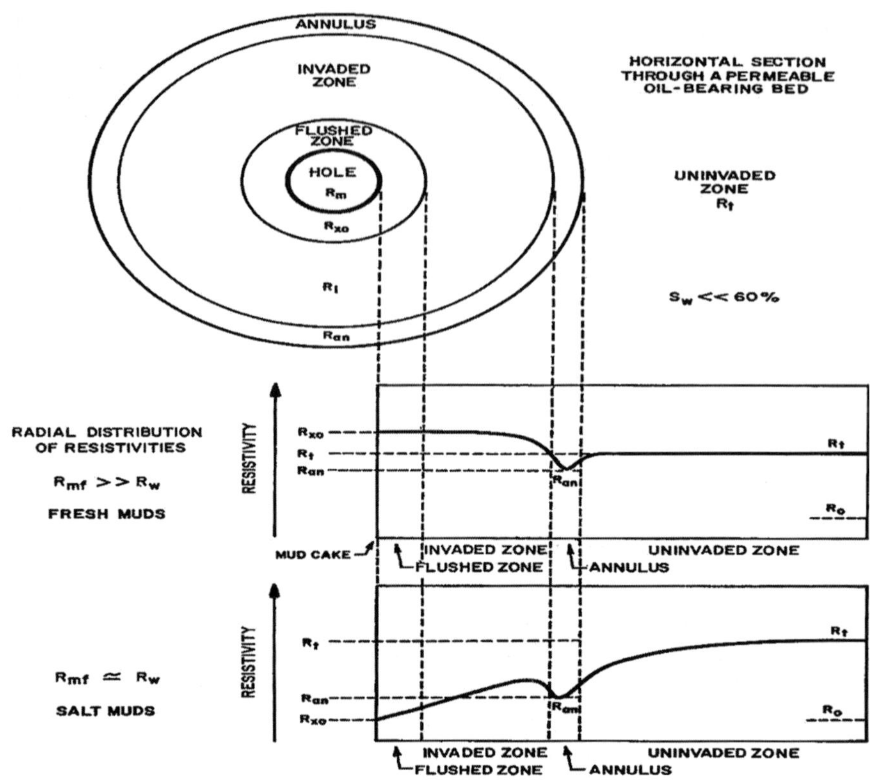

Fig. 5.2 Resistivity Profile-Hydrocarbon Zone (AAPG, 1982)

(1) In **freshwater muds**—the resistivity of both the mud filtrate (R_{mf}) and the residual hydrocarbons (R_h) is much greater than that of the formation water (R_w); therefore, the resistivity profile is built up where the shallow flushed zone (R_{xo}) is comparatively high (the flushed zone has both mud filtrate (R_{xo}) and some residual hydrocarbons (R_h). The **medium invaded zone (R_i)** has a mixture of mud filtrate (R_{mf}), formation water (R_w), and some residual hydrocarbons (R_h). Such a mixture causes high resistivity readings; a hydrocarbon zone invaded with freshwater mud results in a resistivity profile where the shallow (R_{xo}), medium (R_i), and deep (R_t) resistivity tools, all record high resistivities. In some instances, the deep resistivity is higher than the medium resistivity due to the annulus effect. The **uninvaded zone** causes higher resistivity than if the zone had only formation water (R_w), because hydrocarbons are more resistant than formation water (R_w).

So: "$R_t > R_o$"; "R_t"—is normally somewhat less than that of the flushed (R_{xo}) and invaded "R_i" zones. Generally, $R_{xo} > R_i >$ which is more or less than R_t.

(2) In **saltwater muds**—Because the resistivity of the mud filtrate (R_{mf}) is approximately equal to the resitivity of formation water ($R_{mf} \sim R_w$), and the amount of residual hydrocarbons (R_h) is low, then the resistivity of the flushed zone (R_{xo}) is low. As more hydrocarbons mix with mud filtrate in the invaded zone, the resistivity of the invaded zone (R_i) begins to increase.

Generally: $R_t > R_i > R_{xo}$.

5.4 Formation Temperature (T_f)

Formation temperature is important in log analysis because the resistivities of the drilling mud (R_m), the mud filtrate (R_{mf}), and the formation water (R_w) vary with temperature. The formation temperature can be calculated by defining the slope or temperature gradient from the following equation (Schlumberger 1974, Wylie and Rose 1950):

$$m = \frac{y - c}{x} \tag{5.4}$$

where:

m = temperature gradient
y = bottom hole temperature (BHT)
c = surface temperature
x = total depth (TD)

5.5 Specific Log Types

Specific log types such as SP, resistivity, porosity, and gamma ray logs are discussed in detail by Asquith and Gibson (1982).

5.5.1 Spontaneous Potential Logs (SP)

A spontaneous potential log is a record of direct current (DC) voltage differences between the naturally occurring potential of a moveable electrode running down the borehole and that of a fixed electrode located at the surface, measured in millivolts. The SP log, influenced by parameters affecting the borehole environment, is used only with conductive (saltwater-based) drilling muds, to detect permeable beds and their boundaries, and to determine formation water resistivity (R_w) and the volume of shale in permeable beds. The magnitude of SP deflection is due to the difference in resistivity between mud filtrate (R_{mf}) and formation water (R_w), and not to the amount of permeability.

The response of shales is relatively constant and follows a straight line called a shale baseline. SP curve deflections are measured from the shale baseline.

The detection of hydrocarbons by the suppression of the SP response is another use of the SP curve. The presence of shale in a permeable formation reduces the SP deflection. In hydrocarbon-bearing zones, the amount of SP reduction is greater than the volume of shale and is known as "hydrocarbon suppression" (Hilchie 1978). The SP curve can be suppressed by thin beds, shaliness, and the presence of gas. The spontaneous potential curve (Fig. 5.3) shows deflection from the shale baseline, with resistivity of the mud filtrate (R_{mf}) much greater than that of formation water (R_w); at the top of the diagram, the static spontaneous potential (SSP), is the maximum deflection possible in a thick, shale-free, and water-bearing (wet) sandstone for a given ratio of R_{mf}/R_w.

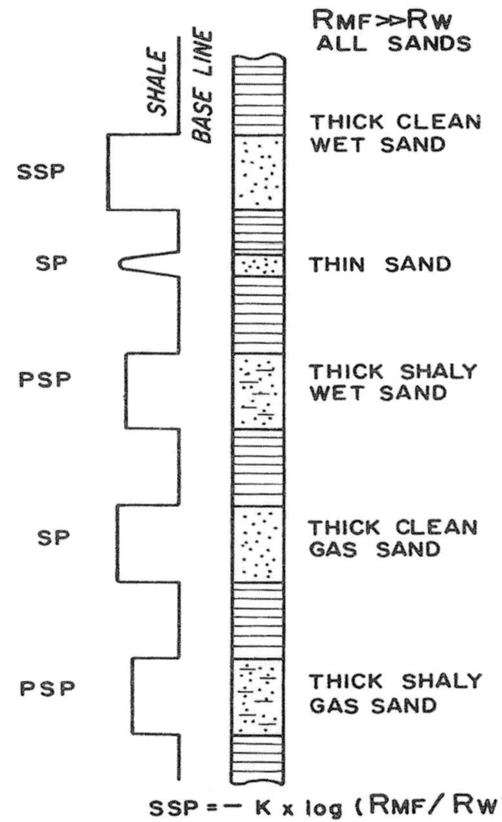

Fig. 5.3 SP deflection with resistivity of the mud filtrate much greater than formation water (Rmf >> Rw) (AAPG, 1982)

SSP is a necessary element for determining accurate values of R_w and volume of shale. In the diagram, SP shows the response of the spontaneous potential curve to the presence of thin beds and/or the presence of gas. Pseudo-static-self spontaneous potential (PSP) is the spontaneous potential curve response if shale is present.

In salt dome drilling, well logging is carried out in salt mud (conductive) or oil mud (nonconductive), and can be accomplished with satisfactory results. Salt muds affect the SP and the resistivity curves on electric logs. With a highly conductive fluid in the borehole, the SP tends to flatten out or may even reverse. In nonconductive or oil base muds, the SP and conventional resistivity (excluding the induction-conductivity curves), are totally ineffective.

5.5.2 Resistivity Logs (R)

A resistivity log is an electric log that can be used to determine resistivity porosity and hydrocarbon versus water-bearing zones, and can define permeable zones. Because the reservoir rock's matrix is nonconducive, the ability of rock to transmit a current is mainly a function of water in the pores. Hydrocarbons, the same as the rock's matrix, are nonconductive; so when the hydrocarbon saturation of the pores increases, the rock's resistivity also increases.

The resistivity log can be used to determine a formation's water saturation (S_w) by applying the Archie equation (Archie 1942):

$$S_w = \frac{(F.R_w)^{1/n}}{R_t} \qquad (5.5)$$

where:

S_w = water saturation,
F = formation factor ($a/\o^m$),
a = turtuosity factor,
m = cementation factor,
R_w = resistivity of formation water,
R_t = true formation resistivity as measured by a deep reading resistivity log,
n = saturation exponent ($\sim$2.0).

There are two basic types of resistivity logs that measure formation resistivity: namely, induction and electrode logs.

5.5.2.1 Induction Electric Devices

There are two types of induction devices: the induction electric log and the dual induction focused log.

The **Induction Electric Log (IEL)** is an electric log that measures conductivity; "or actually true resistivity from conductivity"; it is composed of three curves: short normal, induction, and spontaneous potential (SP). It is normally used with fresh-water drilling mud ($R_{mf} > 3R_w$); but because it does not require the transmission of electricity through drilling fluid (electric current generated by coils that produce an electromagnetic signal which induces currents in the formation), it can therefore be run in air-, oil-, or foam-filled boreholes.

The **Dual Induction Focused Log (DIFL)** is used in freshwater drilling muds and consists of a deep-reading induction device (R_{ILD} that measures R_t) similar to the IEL, a medium-reading induction device

(R_{ILM} measures R_i), and a shallow reading (R_{xo}) focused laterolog (similar to short normal); the shallow reading laterolog may be either a Laterolog-8 (LL-8) or a spherically focused log (SFL). The dual induction focused log is used in formations deeply invaded by mud filtrate; when freshwater drilling muds invade through a hydrocarbon-bearing formation (Sw $<<$ 60%), there is high resistivity in the flushed zone (R_{xo}), the invaded zone (R_i), and the uninvaded zone (R_t).

Figure 5.4 is an example of a hydrocarbon-bearing formation zone; the "DIFL" shown on the right side of the log (Tracks #2 and #3) is a logarithmic scale from 0.2–2000 ohm-meters from left to right. It comprises three resistivity values:

1. **deep induction log resistivity curves (R_{ILD})**, which measure the true resistivity (R_t) of the uninvaded zone (at depths of 8748–8774 ft), where the curves read a high resistivity ($\sim$50), because hydrocarbons are more resistant than the formation salt water ($R_t > R_w$).

2. **Medium induction log resistivity curves (R_{ILM})**, which measure the resistivity of the invaded zone ($R_i = \sim$60); the curve records a high resistivity reading due to a mixture of mud filtrate (R_{mf}), formation water (R_w), and residual hydrocarbon (R_h) in the pores. Such resistivity is equal to or slightly more than the ILD curve; whereas, in the annulus zone, the ILM may record slightly less than the ILD curve.

3. The **spherically focused log resistivity curves (R_{SFL})**, which measure the resitivity of the flushed zone ($R_{xo} = \sim$125); the curve reads a higher resistivity than the deep (ILD) or medium (ILM) induction curves, because the flushed zone contains both the mud filtrate and residual hydrocarbons (Rh). The following ratios are also needed for the tornado chart calculations: $R_{SFL}/R_{ILD} = 3$ and $R_{ILM}/R_{ILD} = 1.2$.

5.5.2.2 Electrode Resistivity Logs

Electrode logs, which are the second type of resistivity measuring devices, are used with salt-saturated drilling muds ($R_{mf} = R_w$); examples of electrode resistivity tools are: normal, lateral, Laterolog, Microlog, Microlaterolog, Proximity Log, and Spherically Focused Log. In salt water drilling muds, when it invades a

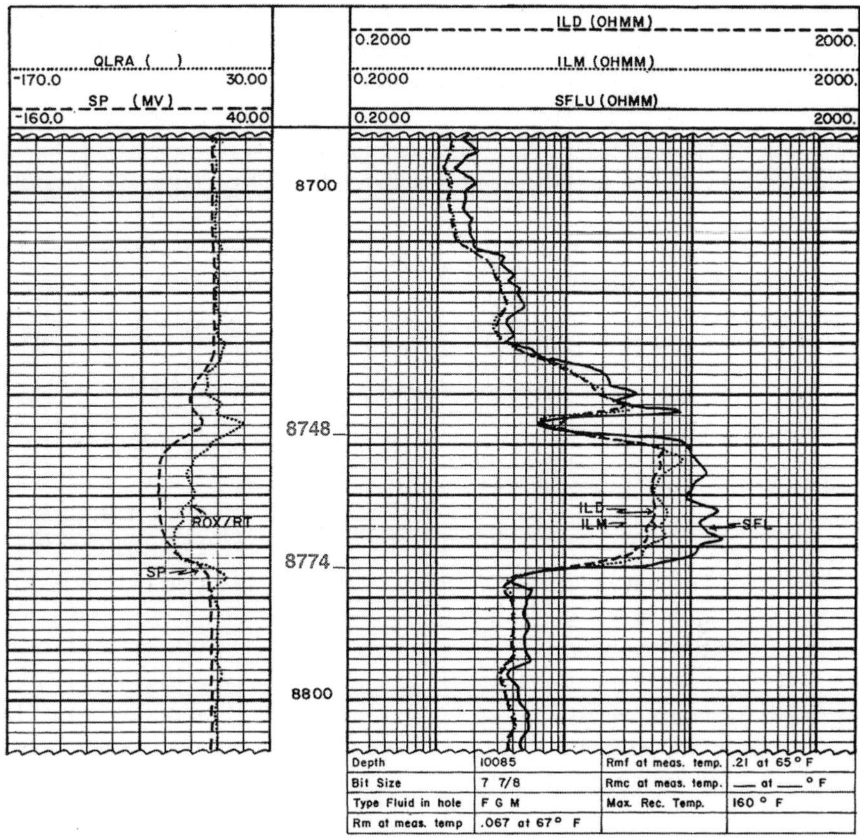

Depth	10085	Rmf at meas. temp.	.21 at 65° F
Bit Size	7 7/8	Rmc at meas. temp.	__ at __ ° F
Type Fluid in hole	F G M	Max. Rec. Temp.	160 ° F
Rm at meas. temp	.067 at 67° F		

Fig. 5.4 Dual induction focused log curves through a hydrocarbon-bearing zone (using freshwater-based drilling mud) (Schlumberger, AAPG, 1982)

hydrocarbon-bearing zone ($S_w << 60\%$), there is low resistivity in the flushed zone (R_{xo}), an intermediate resistivity in the invaded zone (Ri), and high resistivity in the uninvaded zone (R_t); the reason for the increase in resistivities deeper into the formation is the increase of the hydrocarbon saturation (Sh).

The **Laterolog** is designed to measure true formation resistivity (R_t) in boreholes filled with saltwater muds.

The **Dual Laterolog (LLD)—Microspherically Focused Log (MSFL)**—consists of three curves: a deep reading (R_t) resistivity device or the uninvaded zone (R_{LLd}); a shallow reading (R_i) resistivity device (R_{LLs}); and the (R_{SFL}). The Microspherically Focused Log is a pad type, focused electrode log that has a very shallow depth of investigation, and measures resistivity of the flushed zone (R_{XO}). If the MSFL runs together with the Dual Laterolog (R_{LLD} & R_{LLS}), the three curves are displayed in tracks #2 and #3 of the log on a four-cycle logarithmic scale from 0.2 to 2000, values increasing to the right; the three curves are used to correct (for invasion) the deep resistivity (R_{LLD}) to

true formation resistivity (R_t), the diameter of invasion (d_i), and the ratio of R_t/R_{XO}, by the tornado chart (Asquith and Gibson, AAPG 1982).

Figure 5.5 shows the DLL-MSFL curves through a hydrocarbon-bearing zone where the drilling mud is saltwater-based. On the right side of the log (tracks #2 and #3), resistivity values are higher as distance increases:

(a) **Deep Laterolog resistivity curves (R_{LLD})** measure the true resitivity (R_t "dashed lines"), which is that of the uninvaded zone (at a depth 9326 ft, it reads a value of $\sim$18.0); in general, at the interval 9306–9409 ft, the curve reads high resitivity because of high hydrocarbon saturation in the uninvaded zone (R_t).

(b) **Shallow Laterolog resistivity curves (R_{LLS})** measure the resistivity in the invaded zone (R_i). The shallow Laterolog (LLS) records a lower resistivity than the deep Laterolog (LLD) because the invaded zone (R_i at depth 9326 ft reads a value

Fig. 5.5 Dual Laterolog-Microspherically Focused Log (MSFL) curves through hydrocarbon-bearing zone (Saltwater-based drilling mud) (Schlumberger, AAPG, 1982)

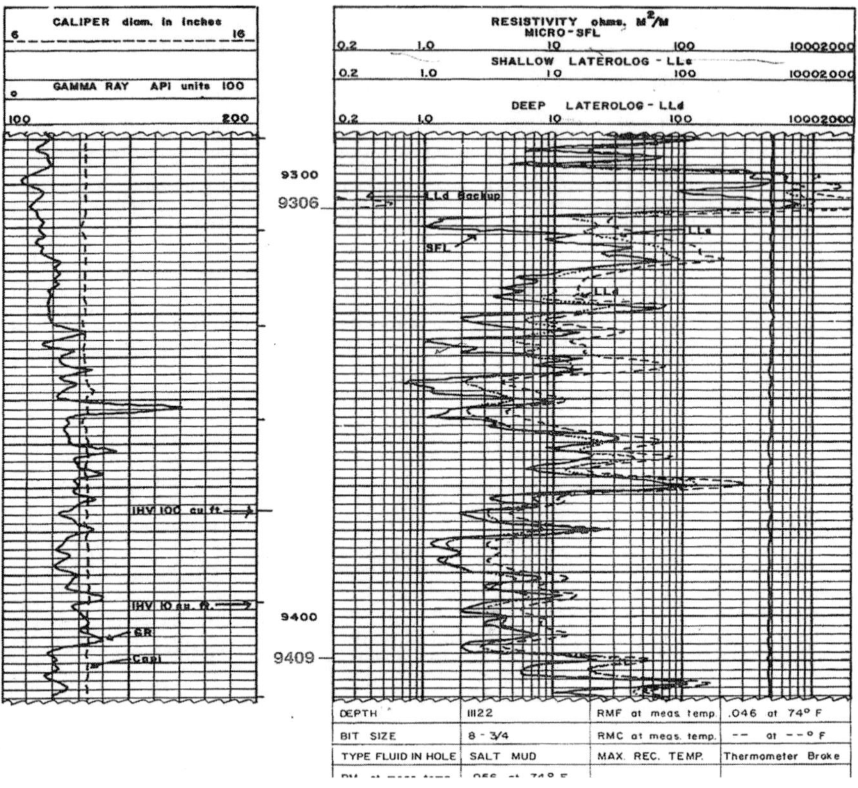

of ∼10) has a lower hydrocarbon saturation (S_h) than the uninvaded zone (R_t).

(c) **Miscrospherically focused log resistivity curves (MSFL)** measure the resistivity of the flushed zone (R_{xo} "solid line"); the curve records low resistivity because of the low resistivity of both the saltwater mud filtrate (R_{mf}) and the residual hydrocarbon saturation (R_h) in the flushed zone (R_{xo}). The uninvaded zone (R_t) has a high resistivity, the invaded zone (R_i) has a lower resistivity, and the flushed zone (R_{xo}) has the lowest resistivity (at depth 9326 ft, it reads a value of ∼3.5).

5.5.2.3 Micrologs (ML)

The Microlog is a pad type resistivity device that primarily detects the mud cake by two resistivity measurements: the micro normal which investigates 3–4 inches (or 7–9 cm) into the formation (R_{xo}), and the micro inverse which investigates 1–2 inches (or 2.35–4.5 cm) and measures the resistivity of the mud cake (R_{mc}). The detection of the mud cake by the ML indicates that

invasion has occurred and the formation is permeable. Permeable zones show up as positive separation when the micro normal curves read higher resistivity than the micro inverse curves. Shale zones show no separation or negative separation (micronormal < microinverse).

The ML does not work well in saltwater-based drilling muds ($R_{mf} \sim R_w$) because the mud cake may not be strong enough to keep the pad away from the formation, and hence positive separation cannot occur.

The Microlaterolog (MLL) and the Proximity log (PL) are designed to measure the resistivity of the flushed zone (R_{xo}). The MLL should run only with saltwater-based drilling mud because it is strongly influenced by mud cake thicknesses greater than $\frac{1}{4}$ inch (∼2/3 cm); (Hilchie 1978).

5.5.3 Porosity Logs

Porosity logs include sonic logs, density logs, and neutron logs. The sonic log records matrix porosity, whereas the nuclear logs (density or neutron) determine the total porosity.

5.5.3.1 Sonic Logs

Sonic logs determine porosity in consolidated sandstones and carbonates with intergranular porosity (grain stones) or intercrystalline porosity (sucrosic dolomites). They measure interval transit time (Δt) in microseconds per foot (μsec/ft). The interval transit time (Δt), which is the reciprocal of the velocity of a compressional sound wave in feet per second, depends on both lithology and porosity. Therefore, a formation's matrix velocity must be known to derive sonic porosity by the following formula (Table 5.2):

$$\emptyset_{sonic} = \frac{\Delta t_{log} - \Delta t_{ma}}{\Delta t_f - \Delta t_{ma}} \quad (5.6)$$

where:

$\emptyset_{sonic}$ = sonic derived porosity,
Δt_{ma} = interval transit time of the matrix,
Δt_{log} = interval transit time of the formation,
Δt_f = interval transit time of the fluid in the well bore (fresh mud = 189; salt mud = 185).

The interval transit time (Δt) of a formation is increased because of the presence of hydrocarbons (i.e., the hydrocarbon effect). The hydrocarbon effect should be corrected, because the sonic derived porosity will be too high otherwise. The following empirical corrections for the hydrocarbon effect were proposed by Hilchie (1978):

$$\emptyset = \emptyset_{sonic} \times 0.7 \text{(gas)}$$
$$\emptyset = \emptyset_{sonic} \times 0.9 \text{(oil)}$$

Table 5.2 Sonic velocities and interval transit times for different matrices used in the Sonic Porosity Formula (after Schlumberger 1972)

	V_{ma}(ft/sec.)	Δt_{ma}(μsec./ft)
Sandstone	18,000–19,500	55.5–51.0
Limestone	21,000–23,000	47.6
Dolomite	23,000–26,000	43.5
Anhydrite	20,000	50.0
Salt	15,000	67.0
Casing (Iron)	17,500	57.0

5.5.3.2 Density Logs

Density is a porosity log that measures the electron density of a formation, and it helps identify evaporite minerals, detect gas-bearing zones, determine hydrocarbon density, and evaluate shaley sand reservoirs and complex lithologies (Schlumberger 1972). The density logging device consists of a medium-energy gamma ray source that emits gamma rays into a formation. Gamma rays collide with electrons in the formation; the collisions result in a loss of energy from the gamma ray particle. Electron density can be related to bulk density (ρb) of a formation in gm/cc.

Formation bulk density "ρ_b" is a function of matrix density, porosity, and the density of the fluid in the pores (salt mud, fresh mud, or hydrocarbons). To determine density porosity by calculation, the matrix density and type of fluid in the borehole must be known. The formula for calculating density porosity (Table 5.3) is:

Table 5.3 Matrix densities of common lithologies used in the Density Porosity Formula (after Schlumberger 1972)

	ρ_{ma} (gm/cc)
Sandstone	2.648
Limestone	2.710
Dolomite	2.876
Anhydrite	2.977
Salt	2.0

$$\emptyset_{den} = \frac{\rho_{ma} - \rho_b}{\rho_{ma} - \rho_f}$$

Where:

$\emptyset_{den}$ = density derived porosity
ρ_{ma} = Matrix density
ρ_b = formation bulk density
ρ_f = fluid density (1.1 salt mud, 1.1 fresh mud, and 0.7 gas)

5.5.3.3 Neutron Logs

Neutron logs are porosity logs that measure the hydrogen ion concentration in a formation. In clean formations (i.e., shale-free) that are saturated with oil or water, the neutron log measures liquid-filled porosity.

Neutrons that are created from a chemical source in the neutron logging tool collide with the nuclei of the formation material, resulting in the loss of some of their energy. Maximum energy loss occurs when the neutron collides with a hydrogen atom, because the hydrogen atom and the neutron are almost equal in mass. Therefore, the maximum amount of energy loss is a function of a formation's hydrogen concentration; accordingly, as the hydrogen is concentrated in the fluid-filled pores, the energy loss can be related to the formation porosity.

Neutron porosity is lower in gas-filled pores than in oil- or water filled-pores because there is less concentration of hydrogen in gas compared to oil or water (the gas effect).

5.5.3.4 Combination Neutron-Density Logs

The combination neutron-density log consists of neutron and density curves which are normally recorded in limestone porosity units, with each division equal to either 2 percent or 3 percent porosity. Besides its use as a porosity log, it is also used to determine lithology and to detect gas-bearing zones.

True porosity can be obtained from the neutron-density curves by reading apparent limestone porosities from the neutron and density curves, and then calculating by using the root mean square formula:

$$\text{\O}_{N-D} = \sqrt{\frac{\text{\O}_N^2 + \text{\O}_D^2}{2}} \qquad (5.7)$$

where:

$\text{\O}_{N-D}$ = neutron-density porosity
$\text{\O}_N$ = neutron porosity (limestone units)
$\text{\O}_D$ = density porosity (Limestone units)

Figure 5.6 is a schematic illustration of the effects of lithology on the combination gamma-ray-neutron-density log; such a relationship provides a powerful tool for petroleum geologists to construct better facies maps. The figure shows the change in neutron density response between an oil- or water-bearing sand and a gas-bearing sand. The oil- or water-bearing sand has a density log reading of 4 porosity units more than the neutron log, in contrast with the gas- bearing sand which has a density reading of up to 10 porosity units more than the neutron log. The increase in density

porosity with the decrease of neutron porosity is known as the "gas effect," which is caused by gas in the pores. Therefore, the effect of gas on the neutron density log is a very important log response to detect gas-bearing zones.

5.5.4 Gamma Ray Logs

Gamma ray logs are lithology logs that measure natural radioactivity in formations and can be used to identify lithologies, correlate zones, and calculate volume of shale. Shale-free sandstones and carbonates have low concentrations of radioactive material and therefore give low gamma ray readings. As shale content increases, the gamma ray log response increases because of the concentration of radioactive material in shale. However, potassium-feldspar, micas, and glauconite produce a high gamma ray response in clean or shale-free sandstones, and in this case a "spectralog" can be run in addition to the gamma ray log, because it breaks the natural radioactivity of a formation into the different types of radioactive material such as potassium, thorium, and uranium. Gamma ray, dipmeter, and induction/dual induction logs are good for both salt water and oil-mud systems.

Volume of Shale Calculation: Because of the radioactivity of shale, the gamma ray logs can be used to calculate the volume of shale in porous reservoirs by first determining the gamma ray index (I_{GR}):

$$I_{GR} = \frac{GR_{log} - GR_{min}}{GR_{max} - GR_{min}} \qquad (5.8)$$

Some log analysts restrict the use of this formula to gas-bearing formations and use instead:

$$\text{\O}_{N-D} = (\text{\O}_N + \text{\O}_D)\,/2$$

where:

I_{GR} = gamma ray index,
GR_{Log} = gamma ray reading of formation,
GR_{min} = minimum gamma ray (clear sand or carbonate),
GR_{max} = maximum gamma ray (shale).

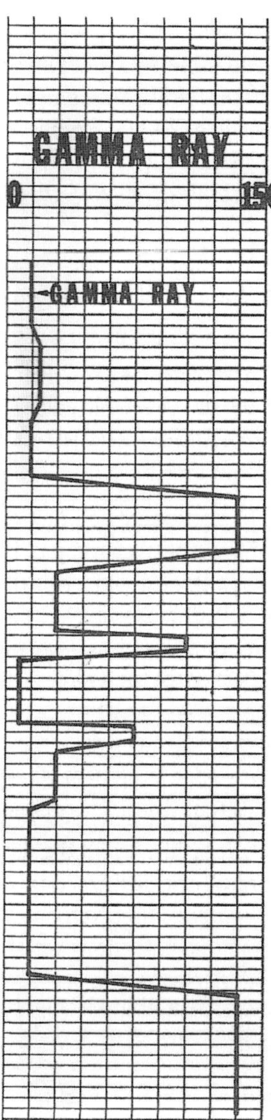
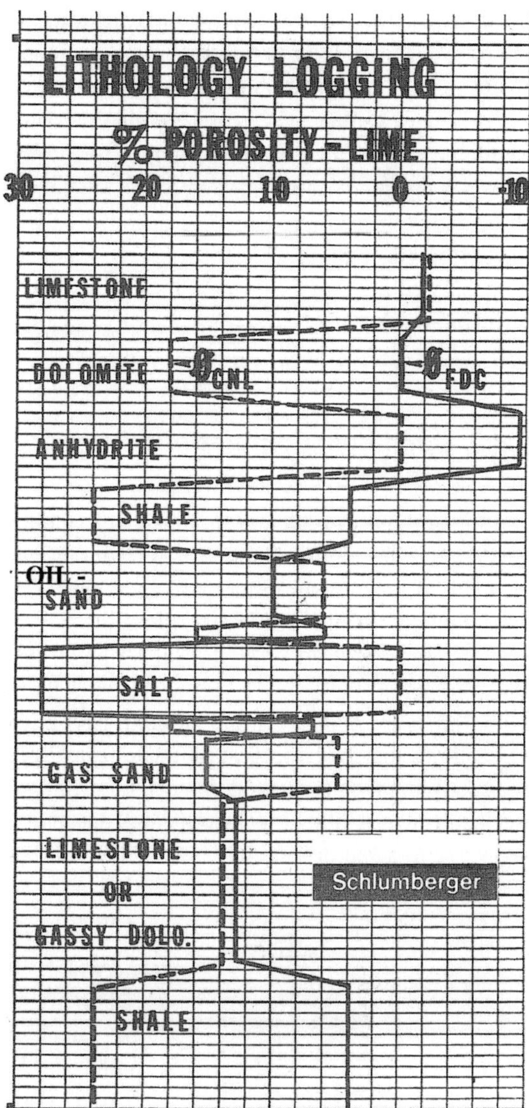

Fig. 5.6 A genaralized lithology logging with combination Gama Ray Neutron-Density Log. It shows relationships between log responsees and the rock type as well as changes from oil-or water-bearing rock units compared to gas-bearing units. (Schlumberger, AAPG, 1982)

The volume of shale is also calculated from the gamma ray index (IGR) by the following Dresser Atlas (1979) formula.

For consolidated/older rocks:

$$V_{sh} = 0.33[2^{(2 \times IGR)} - 1.0] \tag{5.9}$$

For unconsolidated/Tertiary rocks:

$$\text{Volume of shale } (V_{sh}) = 0.083[2^{(3.7 \times IGR)} - 1.0)] \tag{5.10}$$

5.6 Well Design and Well Type Completions

5.6.1 Scope

It is necessary to run casing at various depth intervals and cement it in place at every interval. The number and size of the casing strings vary with the location of the well, its depth, and the anticipated producing characteristics of the objective reservoir. The casing which includes several strings is the thin-walled steel pipe used to line the borehole, to prevent its caving and contamination of potable water reservoirs; to provide anchorage for the wellhead and blowout preventers; and to provide an adequate means of controlling well pressures. Initial cementing jobs are performed in conjunction with setting various casing strings to afford additional support for the casing and to retard corrosion by minimizing the friction factor between the pipe and the corrosive formation waters. Besides the casings, a smaller diameter string known as tubing is added and is considered the actual flow conduit for the produced petroleum.

5.6.2 Open Hole Completions

The open hole completion is the only applicable method for the highly competent formations and is common in low-pressure limestone areas where cable tools are used for the drilling-in operation (Fig. 5.7). Rotary tools are used until the oil string is set; at that time, the cable tool rig moves in, bails the mud from the hole, and drills the desired pay interval. This allows the zone in question to be tested as it is drilled, eliminating formation damage by drilling mud and cement, and allowing incremental deepening as necessary to avoid drilling into water; the latter factor

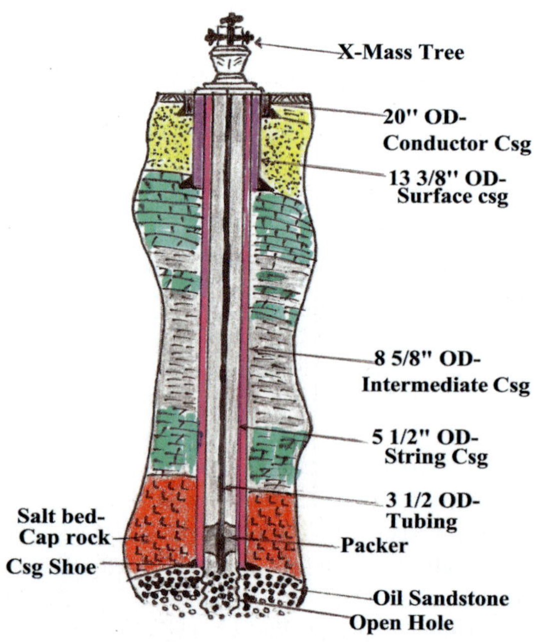

Fig. 5.7 A sketch diagram of an open hole completion in a productive oil well

is quite important in thin, water-drive pay sections consisting of a few feet of oil zone.

In open hole completion, there are three separate casing sizes: the surface pipe, the intermediate, and the oil strings; the oil string is set up at the top of the oil sand in open hole completion, whereas the packer isolates the tubing-oil string annulus from the pay sand. This category refers to wells that flow naturally and is only applicable to highly competent formations with no caving problems.

5.6.3 Perforated Completions

The perforated completion offers a much higher degree of control over the pay section and is only applicable on wells where the oil string is set through the pay section and cemented, to be perforated at the desired interval. **Mechanical perforators** were used prior to the early 1930s, in which casing can be perforated in place; each consists of either a single blade or wheel-type knife, opened at the desired interval, which cuts vertical slots in the casing. **Bullet perforators** are multibarreled firearms designed for being lowered into the well, positioned at the desired interval, and electrically fired from surface controllers. A **jet perforator** uses a steam jet's high velocity impact, which is developed to penetrate the target, causing the liner's inward collapse and partial disintegration.

5.6.4 Screening Techniques

The completion of a well in unconsolidated sand is not simple, because of the problem of excluding sand produced together with the oil; unchecked sand can cause corrosion of equipment, and plugging of flow in the string to the extent that the well operation becomes uneconomical. Sand production is also sensitive to the rate of fluid production, especially when large quantities of sand are carried along in the production stream.

Screening techniques are employed for excluding sand by one of two methods: either using slotted (or screen) liners, or by packing the hole with aggregates of gravel. The basic requirement in both methods is that the openings through which the produced fluids flow should be of the proper size to cause the formation sand to form a stable bridge and thereby be excluded.

5.6.4.1 Screen Liners

The slotted or screen liner is normally run on tubing and hung inside the oil string opposite the producing zone. The oil string may have been either cemented through the section and perforated, or set on top of the pay zone (in open hole completion). The maximum opening size that will exclude a given sand is determined from screen analysis. The appropriate liner slot or screen width is taken as twice the 10 percentile size given from screen analysis (Tausch and Corley 1958). Highly variable sands may have the same 10 percentile size, and smaller slots or screen openings are advisable if an unususaly high percentage of fine grains are present (Fig. 5.8). It is extremely important that the sand face be free of mud cake before the liner is set, to prevent plugging; that can be accomplished either by using clay-free completion fluids or by washing the well with salt water before hanging the liner.

5.6.4.2 Gravel Packing

Gravel packing is performed in either perforated or open hole intervals. The gravel size for gravel packs is about five times the 10 percentile sand size. Since the screen is used to exclude gravel only, the slots may be slightly smaller than the gravel; the required thickness of the gravel pack is only four or five gravel diameters. The formation sand then bridges within the pores of the gravel pack, while gravel entry is prevented by the screened liner (Fig. 5.9). In general, open borehole completion is initially cheaper, but the perforated

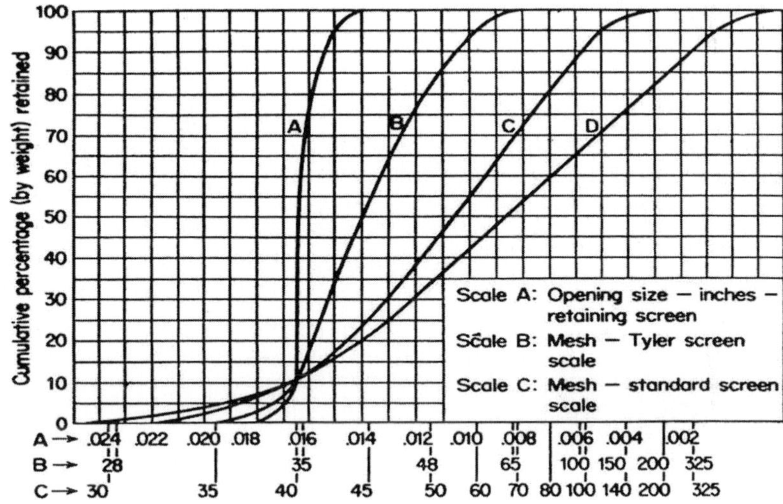

Fig. 5.8 Method of plotting screen analysis data. Note that four sands shown have ossentially same 10 percentile point of 0.0165-in. despite size range variations. After Tausch and Corley,[25] courtesy *Petroleum Engineer*

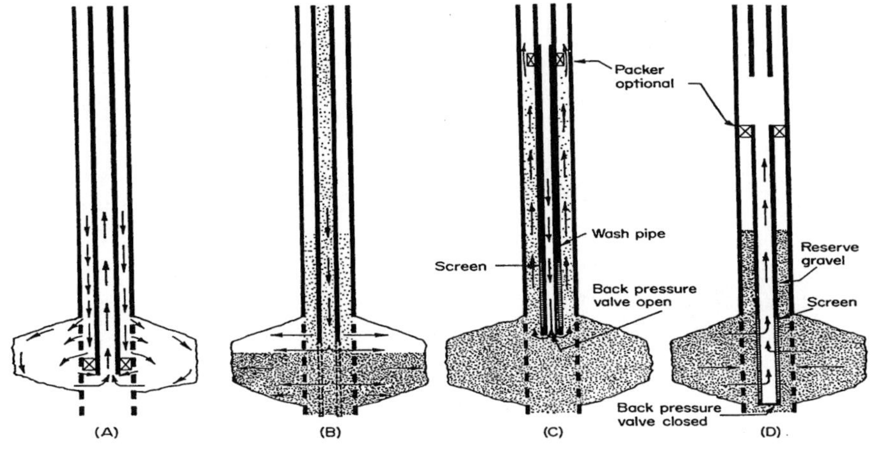

Fig. 5.9 Simplified method of gravel packing commonly used in the Gulf Coast: (**A**) Perforations are washed; (**B**) gravel is squeezed through perforations; (**C**) cavity is filled and screen is washed through gravel; (**D**) after screen placement, wash pipe is removed. After Tausch and Corley,[25] courtesy *Petroleum Engineer*

well completion by hydraulic fracturing offers a much higher degree of control over the pay section.

5.6.5 Other Type Well Completions

A permanent-type well completion is one in which the tubing is run and the wellhead is assembled only once in the life of the well. Perforating, swabbing, squeeze cementing, gravel packing, or other completion or remedial work is performed with special small diameter tools designed to run inside the tubing.

In multiple zone completions, there are one or more separate pay zones producing simultaneously from the same borehole, without commingling of the fluids; this segregation is commonly required for purposes of reservoir control under state regulatory bodies. Dual completions are the most common, although triple and quadruple zone completions have been performed (Gatlin 1960).

References

Archie GE (1942) The electrical resistivity log as an aid in determining some reservoir characteristics. *J. Pet. Technol.*, 5:54–62

Asquith GB, Gibson CR (1982) Basic well log analysis for geologists. The American Association of Petroleum Geologists, Tulsa, Oklahoma, Fig 1/p7; Fig 5/p14; Fig 7A/p20; Fig 7B/p22; Fig 10B/p30.

Dresser Industries, Inc. (1975, 1979) Dresser Atlas Log interpretation fundamentals and charts. Houston, Texas, 107 pp

Hilchie DW (1978) Applied openhole log interpretation. Golden, Colorado, D.W., Hilchie, Inc., in AAPG (1979)

Schlumberger (1972) Log interpretation manual/principles. Schlumberger Well Series, Inc., Ridgefield, Vol. I

Schlumberger (1974) Log interpretation manual/applications. Schlumberger Well Series, Inc., Ridgefield, Vol. II.

Schlumberger (1977) Log Interpretation/Charts. Houston, Schlumberger, Well Sevices, Inc.

Schlumberger (1979) Log Interpretation/Charts. Houston, Schlumberger, Well Sevices, Inc.

US Environmental Protection Agency (1977) An introduction to the technology of subsurface waste injection. Environmental Protection Agency 600/2-77-240, 1977; Cincinnati, Ohio, pp 73–88

Wylie MRJ; Rose WD (1950) Some theoretical considerations related to the quantitative evaluation of electric log data. Jour. Petroleum Technology 189:105–110

Tausch GH and Corley CB Jr (1958) Sand exclusion in oil and gas wells. The Petroleum Engineer, ppB-38, and B-58, respectively, in: Gatlin C (1960) Petroleum engineering, drilling and well completions. Dept. of Petroleum Engineering, Fig 15.9/p314 & Fig 15.12/p315, Univ. Texas, Prentice–Hall, Inc., Englewood Cliffs, NJ USA

Part II
A Guide to Lithiostratigraphic Correlation Charts – A Model Type/North Africa

Chapter 6

A Lithostratigraphic Correlation Chart—A Model Type/North Africa

A Case Study

6.1 Introduction

6.1.1 Aspect

The northern African countries include Egypt in the east, Libya to the west, and the Arabian Maghreb to the northwest, which comprises Morocco, Algeria, and Tunisia (Fig. 6.1; Michel, 1969, AAPG 1970).

During the 18th century, the French colonialists spread widely all over the Arabian Maghreb countries, where Algeria, in 1962 the last French colony to gain its independence, is bordered by Morocco and Mauritania on the west, Tunisia and Libya on the east.

Algeria is a country of ~2.36 million km^2 (900,000 square miles), with a population of about 35 million; the Algerian territory covers an area between latitude 20°–36°N and longitude 10°E–08°W, which is approximately equal to that of its neighbors combined.

Libya is a country of 1.78 million km^2 (680,000 square miles) and was considered almost barren by an international mission after the Second World War; but prompted by discoveries in neighboring Algeria, seventeen companies carried out oil exploration and had invested more than $800 million by the beginning of 1963.

Egypt is a country of 1.21 million km^2 (460,000 square miles), with a population of about 75 million. Egypt has been called "the gift of the Nile," yet the Nile Valley is one of the poorest regions in North Africa, although holding mineral resources such as iron phosphates, gypsum, salt, and potash.

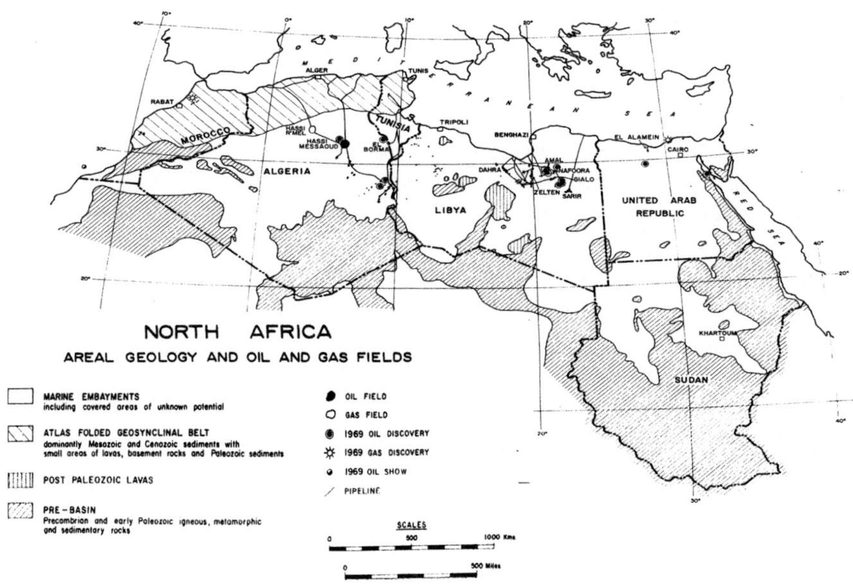

Fig. 6.1 North africa geology and oil and gas fields. (Michel 1969, AAPG 1970)

F.A. Assaad, *Field Methods for Petroleum Geologists*,
© Springer-Verlag Berlin Heidelberg 2009

6.1.2 Petroleum Perspective

The sedimentary formation in the Arabian Maghreb accumulated from the Tethys Ocean, which flooded the area during the period between the Cambrian and the Eocene. From the Precambrian window of the Hoggar Mountains southeast of Algeria towards the Mediterranean, successively younger sediments accumulated in the epicontinental zones. Consequently, younger oil fields appear from the southwest towards the northeast. The traps, irrespective of age, were frequently controlled by Precambrian-Paleozoic structures.

The Arabian Maghreb is well known as the continent's largest oil, gas, phosphate, and lead supplier. Egypt has limited mineral resources that are concentrated in the oases of the western and eastern deserts and the Sinai. After independence, the National Algerian Petroleum Society (Société Nationale de Transport et de Hydrocarbures, Sonatrach), which is the Algerian state petroleum company, increased its size with the French state's Sopefal, whereas Libya increased its oil production significantly after the military revolution in 1969. In the same period, Tunisia and Morocco awarded large blocks of their offshore acreage to French companies; and Russian and American oil companies discovered gas in the Nile Delta and oil in the northwestern zone of Egypt.

In 1965, crude oil reserves of the Arabian peninsula and North African countries represented almost two-thirds of the world total (300 BBO). The African rate of expansion was tens of times as high as the overall increase in world production in the past three decades.

6.1.2.1 The Algerian Territory

During the colonial epoch, French oil companies Sopefal and Becip carried out petroleum exploration studies in north Algeria, whereas other French petroleum companies such as SN-Repal and Total were granted long-term contracts to carry out production operations in different basins of the Saharan Platform to the south. These companies worked independently; geologists and research scientists who joined Sonatrach later, after independence, found it difficult to review hundreds of documents and publications in different basins; numerous geological terms and confusing abbreviations for the same lithological and/or stratigraphical series caused identical problems for defining certain stratigraphic levels on a regional scale.

There are four main regions of mineral resources in the Algerian Sahara: the Precambrian mass of the Hoggar which constitutes the anchor of the Sahara (rich in platinum, asbestos, tin, zinc, etc.); the Paleozoic to the south; the Triassic Province to the north; and the Eocene to the east (Kun 1965; Fig. 6.2a).

The Algerian Sahara can be classified into four groups of structures and basins: (1) the gas field of Hassi er R'Mel, 320 km (200 miles) south of Algiers, assigned to SN Repal and CFP, which is the world's largest gas field, and which had been expanding rapidly since the late 1960s; (2) the oil field of Hassi Messaoud, 300 miles south of Constantine, owned by SN Repal; (3) the Salah Basin (480 km or 300 miles south of Hassi er R'Mel), where oil was struck at Edjelah (CREPS); and (4) the Fort Polignac-Tinrhert Basin, the largest oil producer in Africa until 1963, which is owned by CREPS (Kun 1965; Fig. 6.2b), and consists of the Zarzaitine, Tiguientourine, and Edjeleh fields and a number of others. The Zarzaitine field comprises different oil stratigraphic levels in the Devonian formation. The Fort Polignac Basin lies near the western Libyan border where the northeastern part has been intensely faulted by the Mesozoic and post-Jurassic rejuvenation of Paleozoic movements. In that sector, the faulted Ohanet anticline trends north-south on the Tinrhert Plateau. At Ohanet, oil accumulates in Lower Devonian sandstones at a depth of 2,343.75 m (7,500 ft) The structure of the southern fields is very different from the flat-lying basins of Hassi Messaoud and Hassi er R'Mel. The upper Devonian is the most favorable oil reservoir in the Edjeleh field, where the Edjeleh anticline is oriented north-south (Sonatrach 1972).

The **Triassic Province of Algeria** divides the Northern Sahara into eastern and western basins where Paleozoic strata form a north-south trending ridge along the longitude of Algiers; sediments transgressed the northeastern Sahara and northern Tripolitania from the Permian up to the Triassic period.

6.1.2.2 The Libyan Territory

The Paleozoic Sea covered most of the Libyan territory; in the eastern part, oil fields are oriented east-southeast parallel to Es-Sidr, the Bay of Syrte. Oil and gas shows are found in a broad zone of northwestern Libya, extending from west of Tripoli to Edjeleh and including the Oued Chebbi gas field in the

Fig. 6.2 (a) The mineral resources of the Sahara (Kun, 1965); (b) Fort Polignac oil field (Kun, 1965); (c) The East Libyan oil field (Kun, 1965); (d) The mineral resources of Egypt (Kun, 1965)

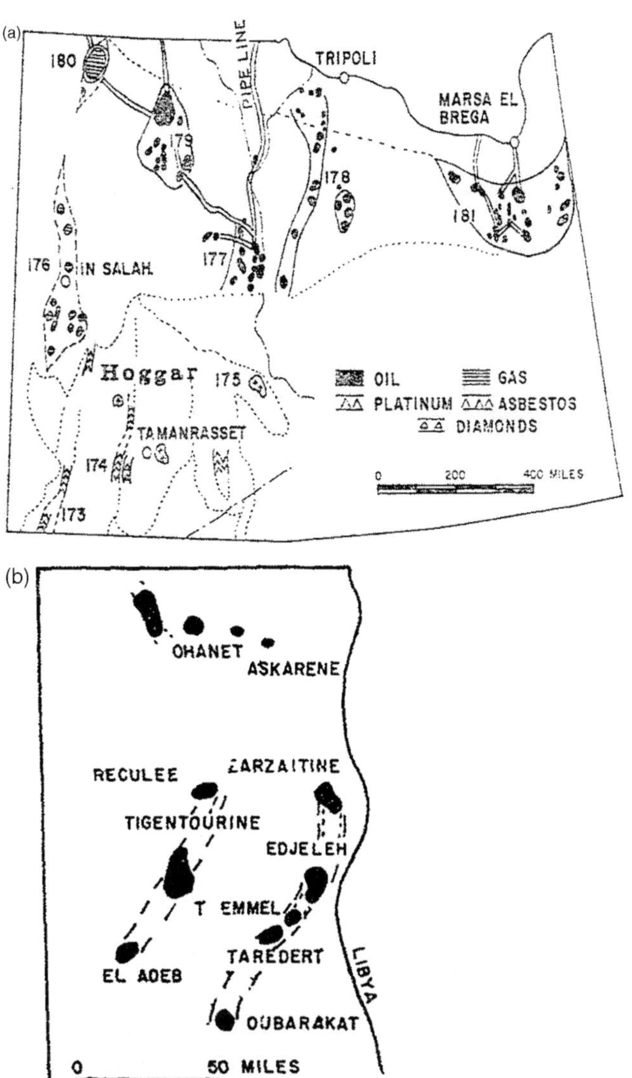

northwestern corner of Tripolitania and Bir el-Rhezeil, yielding more than 2 million ft^3 of gas (Fig. 6.2c; Kun 1965). In eastern Libya, the Paleocene is particularly saline, but thick interbeds of dolomite and anhydrite were deposited during the Lower Eocene. In the Middle Eocene, the subsidence continued, as shown by limestone and thin seams of interbedded lignite and coal. In the Upper Eocene another emergent phase which reached its climax in the Oligocene was initiated, as indicated by conglomerates and sands.

The basins of Tripolitania and northern Fezzan extend along a south-southwest trend for 640 km (400 miles) as a long broad arc. However, the Oued Chebbi gas field, at the northernmost part of the alignment, belongs to the northern Triassic basin.

As in Algeria, the prebasin relief of Libya strongly affects the accumulation of hydrocarbons. Sediments were first deposited in the troughs, and then over the whole pre-Cretaceous land surface. The differential compaction of sediments along with tectonics further impressed the imprint of the Paleozoic-Precambrian topography. Upper Cretaceous carbonates form more favorable traps than Lower Cretaceous sandstone, although basal sandstone and partly dolomitized calcarenites of the coast constitute the principal reservoirs. As in the northern Sahara, many of these reservoirs are located on uplifts, structural heights, or their edges. The Samah and Waha fields are good anomalies of Upper Cretaceous accumulation of oil (Kun 1965).

Fig. 6.2 (continued)

(c)

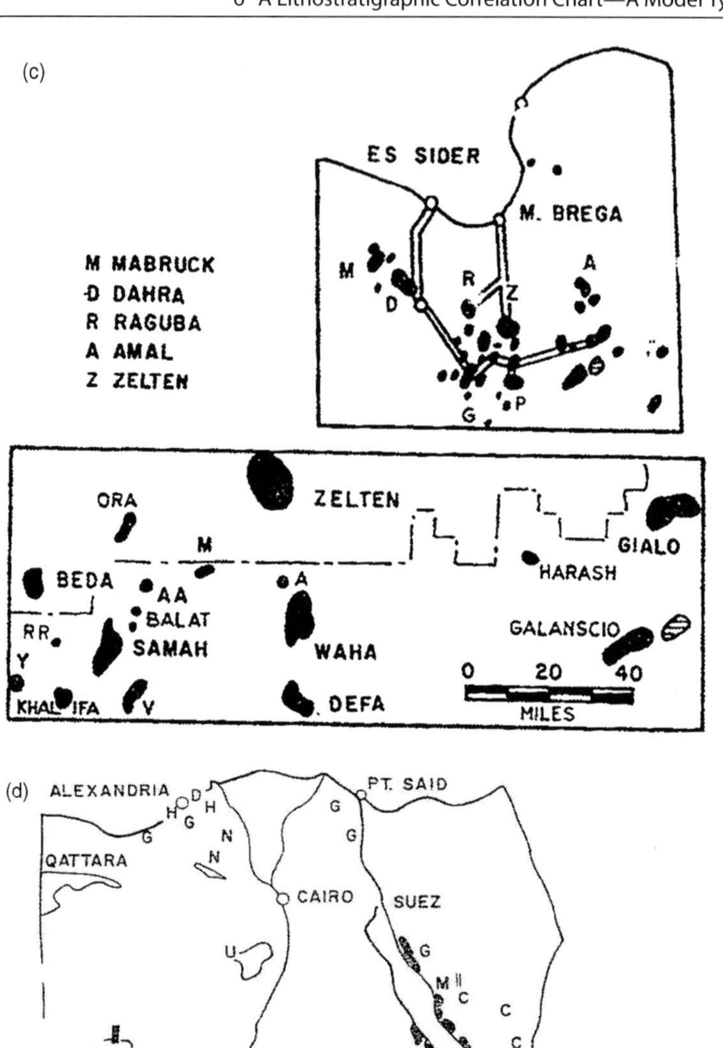

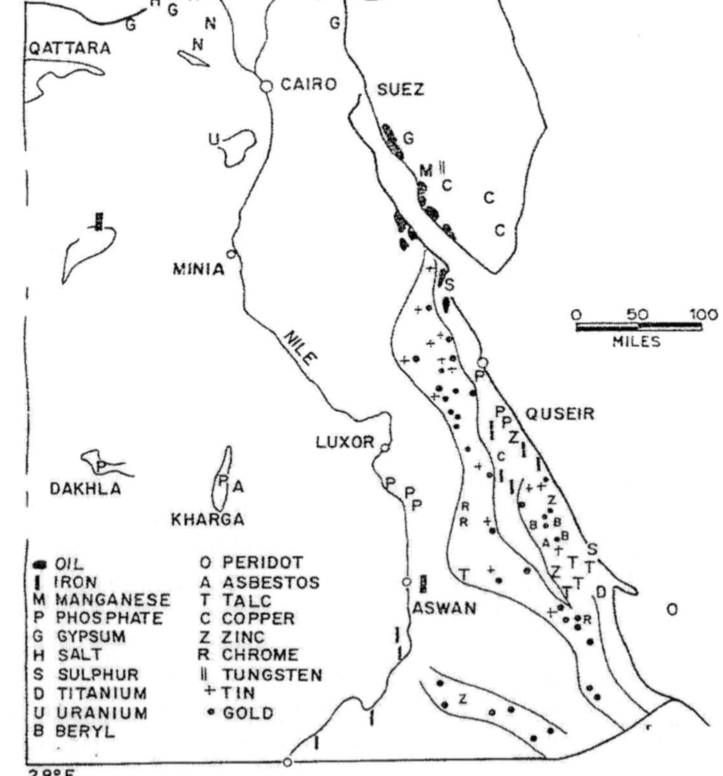

6.1.2.3 The Egyptian Territory

In 1963, oil production was concentrated in the east of the country in Sinai and on the southern coastal areas of the Gulf of Suez (Fig. 6.2d); gas was recently explored in the Nile Delta of Egypt. The Miocene Gulf of Suez oil fields are in fact small coastal basins, whereas the Egyptian Sahara has yielded little oil until the present. The rest of the Sahara can be divided into three unequal structural units, the east, the center, and the west, and can be also classified into three stratigraphic units, the Eocene, Triassic, and Paleocene basins.

6.2 Geology of Algeria

6.2.1 Tectonic History

North Africa is a single depositional basin on the northern shelf of the African craton. The basin generally deepened northward where deposition and marine influence are greater. There is a general conformity of structure throughout most of the Paleozoic until the Hercynian, which marks the collision between Laurasia and Gondwana and which caused regional uplift, folding, and erosion. Paleozoic basins, delineated by earlier tectonic events, were modified, resulting in the development of several intracratonic sag and foreland basins. Petroleum hydrocarbons were generated during the Carboniferous Period within deeper portions of the basins, but uplift normally caused petroleum generation to cease, because subsequent erosion probably removed or dispersed the petroleum hydrocarbons that had accumulated in some areas (Klett, USGS 2000).

Several transgressive-regressive cycles occurred throughout the Paleozoic; two major flooding periods, one in the Silurian and the other in the Late Devonian, were responsible for the deposition of source rocks. Many of the prograding fluvial, estuarine, deltaic, and shallow marine sands that were deposited during these cycles became reservoirs.

During the early Mesozoic, extensional movements caused by the opening of the Tethys and Atlantic oceans developed a cratonic sag basin known as the Triassic Basin. Triassic fluvial sands followed by a thick Triassic to Jurassic evaporite section, were deposited within the sag basin. These sandstones became major reservoirs, whereas evaporites provided a regional seal on top of both the fluvial sandstone and the Paleozoic reservoirs.

Compressional movements during the Late Cretaceous and Pyrenean deformation tilted the Triassic Basin to its present configuration, whereas the existing basins on which the present-day Atlas Mountains lie were inverted by these events (Klett, USGS 2000).

6.2.2 Structural Geology of Algeria

Algeria is subdivided into two major structural units, the northern Algerian Alpine and the southern Saharan Algerian Platform, separated by the south Atlas fault.

6.2.2.1 The Algerian Alpine

The northern Algerian, mainly affected by Alpine orogeny, was formed of young mountains during Tertiary times and consists of a number of structural sedimentary units. From north to south: (**a**) The **off-shore Algerian** is a reduced continental shelf of Tertiary and Quaternary sediments overlying a metamorphic basement. (**b**) The **Tellian Atlas** is a northern complex area comprising a chain of nappes set in place during the Lower Miocene. (**c**) The **Hodna Basin** is a foredeep basin of continental Eocene and Oligocene deposits, overlain by marine Miocene sediments. (**d**) The **High Plateaus** of delineated features in the center with large plains to the west and highs to the east are the foreland of the Alpine range of thin sedimentary cover. (**e**) The **Saharan Atlas** is formed from an elongated trough pinched between the high plateaus and the Saharan platform. The trough was filled in by thick Mesozoic sediments (7,000–9,000 m); later, the Tertiary compressive tectonic stresses modified the former extension trough into a number of reverse structures that led to the creation of the mountain range.

Both the high plateaus and the Saharan Atlas comprise the first major tectonic unit of the Epihercynian platform, which was active during the Alpine stage and caused the separation of the southern portion of the Saharan Atlas from the Saharan Platform, and which constitutes the second major tectonic unit in northeast central Algeria. Both the Epihercynian and Precambrian platforms attained their main tectonic features after the Early Alpine orogenic movements, whereas the successive southerly migrations of the equator during Late Triassic–Early Liassic time led to the formation of several lagoons.

6.2.2.2 The Algerian Saharan Platform

The Algerian Saharan Platform is located to the south of the Alpine domain and is part of the North African craton that yields most of Algeria's hydrocarbon resources. It has been a relatively stable area of low tectonic activity over the course of geologic time. Numerous basement uplifts accompanied by faults have affected the Paleozoic sediments; but the tectonics never reached the violence of the Alpine folds that disturbed the northern portion of the shield and produced folding sheets in the Saharan Atlas Mountains, disengaging the Mesozoic sediments from the plastic saliferous beds of the Triassic/Liassic Formation. No Triassic/Liassic diapirs and no Triassic outcrops were known in the Saharan Platform because of the slow effect of the epeirogenic movements on the relatively thin sediments overlying the solid and hard Precambrian basement (Fig. 6.3).

The Precambrian basement is unconformably overlain by thick sediments, structured during the Paleozoic into a number of separated basins by high zones, followed by Mesozoic and Cenozoic sediments. The basins, from west to east, are: the **Tindouf/Reggane Basins**; the **Bechar Basin**; the **Ahnet-Timimoun Basin**; and the **Mouydir** and

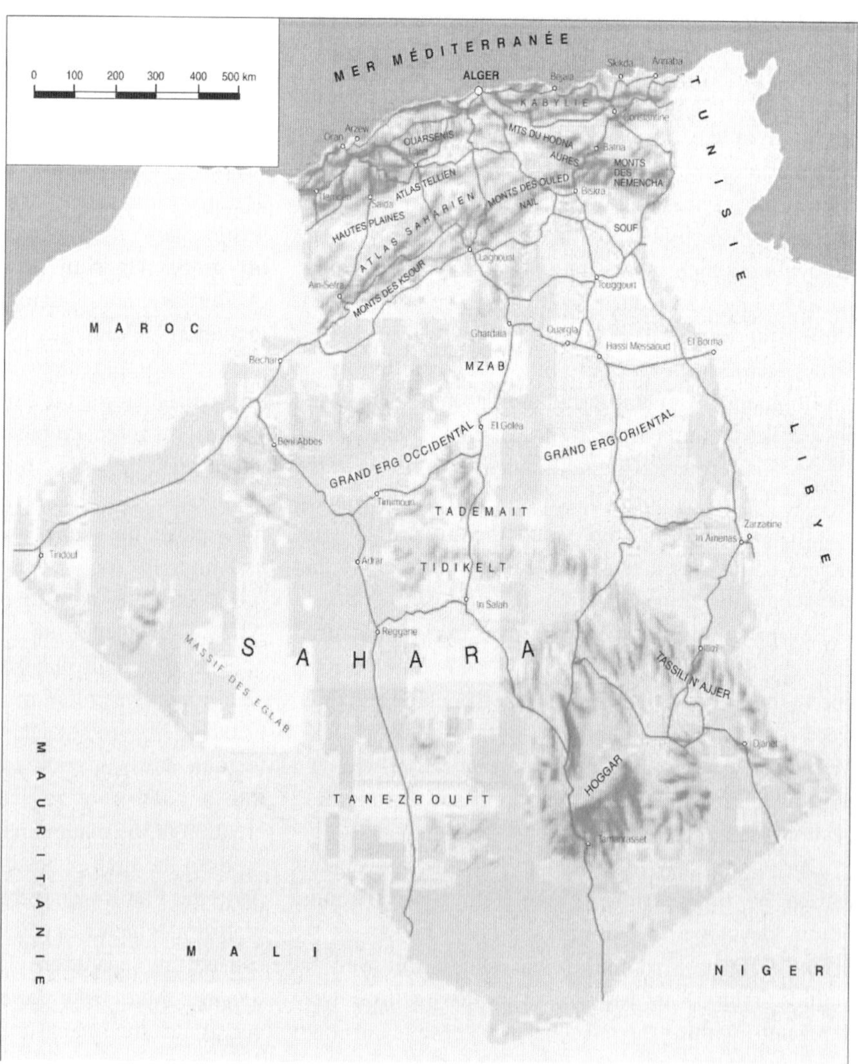

Fig. 6.3 Natural boundaries of algiers and morphological zones http://www. mem-algeria.org

Aguemour-Oued M'ya basins, where major fields are located within the Cambrian of Hassi Messaoud and the Triassic formations of Hassi R'Mel. To the far east, the **Illizi-Ghadames syncline** yields hydrocarbons both in Paleozoic and Triassic sediments.

http://www.mem-algeria.org/hydrocarbons/geology.htm

The Paleozoic structural basins southwest of the Algerian Sahara comprise the following:

1 – **Tendouf Basin** is characterized by an east/west synclinal axis and divides the basin into two unequal parts; a northern structural zone and a southern monocline dipping to the north. Stratigraphically, the basin is represented by a series of Paleozoic formations that thickened in the northern part, pinching to the south, and ranging from Cambro-Ordovician to Upper Carboniferous. The Ordovician and/or Devonian formations played an important role as gas reservoirs.

2 – **Bechar Basin** shows complicated tectonics and stratigraphic aspects of Paleozoic formations (Cambrian to Carboniferous). The granular and the carbonate reservoirs attained poor permeability and fairly good porosity, though secondary fissures especially in the highly tectonic zones might improve the petrophysical characteristics.

3 – **Reggane and the Timmimoun Basins** show gas reservoirs in the Upper Ordovician (Ashgillian), Upper Devonian (Strunian and Frasinian), Lower Devonian (Emisian and Siegenian), and Lower Carboniferous (Visean) sandstones. To the South, the Ahnet and Tinguentour basins are gas producers in the Lower Ordovician, Devonian, and Carboniferous reservoirs. The Ahnet Basin includes three regional north-south trending faults, composed of local isolated highs, and is productive in the Ordovician; whereas the Tinguentour field is productive in the Tournasian formation.

4 – **The Illizi Basin** to the southeast of the Algerian Saharan platform consists of the Tinfoyé Tabenkurt oil and gas fields that are productive from Gothlandian (Unit A) and Upper Ordovician (Unit IV) reservoirs. The Illizi basin is mainly oil and gas productive in the Upper Shaley sandstone series ($T1 + T2$), in the lower sandstone series (SI-series) of the Triassic reservoir (Fig. 6.4), and in the Devonian (Unit $F_3 + F_6$) reservoir. To the south, the Illizi central region is productive from Lower Devonian and Gothlandian reservoir rocks.

6.2.3 Geologic Provinces of the Algerian Sahara

The world is divided into 8 regions and 937 geologic provinces which have been ranked according to the discovered oil and gas volumes (Klett et al. 1997). A geologic province is an area with characteristic dimensions of hundreds of kilometers that encompasses a natural geologic entity (e.g., a sedimentary basin, thrust belt, or accreted terrain) or some combination of contiguous geologic entities. Each geologic province is a spatial entity with common geologic attributes. Province boundaries were drawn as logically as possible along natural geologic boundaries, although in some cases, they were located arbitrarily (e.g., along specific water depth contours in the open oceans). Particular emphasis is placed on the similarities of petroleum fluids within total petroleum systems—unlike geologic provinces, in which similarities of rocks are emphasized.

Klett (2006) prepared a report as a part of the U.S. Geological Survey World Petroleum Assessment 2000 (USGS-WPA 2000), to assess the quantities of conventional oil, gas, and natural gas liquids and to develop a hierarchical scheme of geographic and geologic units. The scheme consists of regions, geological provinces, total petroleum systems, and assessment units. Assessment of undiscovered resources was carried out at the level of total petroleum system or assessment unit. A total petroleum system may be subdivided into two or more assessment units and each assessment unit is sufficiently homogeneous to assess individually in terms of geology, exploration considerations, and risk.

According to USGS-WPA 2000, the total petroleum system includes all genetically related petroleum that occurs in shows and accumulations (discovered and undiscovered), generated by a pod or by closely related pods of mature source rock. Total petroleum systems exist within a limited mappable geologic space, together with the essential mappable geologic elements (source, reservoir, seal, and overburden rocks). An assessment unit is a mappable part of a total petroleum system in which discovered and undiscovered oil and gas fields constitute a single relatively homogeneous population such that the methodology of resource assessment based on estimation of the number and sizes of undiscovered fields is applicable.

A number of factors such as formation thickness (1,000–8,000 m), lithology, tectonic deformations, and

Fig. 6.4 Old and recent classifications of lower liassic/triassic formations in ne central zone of algiers

OLD CLASSIFICATION OF TRIASSIC FORMATIONS	LITHOLOGY	UPDATED CLASSIFICATION OF TRIASSIC DEPOSITS AFTER PALYNOLOGICAL STUDIES		
HORIZON B		LOWER LIASSIC SERIES	Hettangean	INFRA-LIAS
TS1+TS2				
TS3				
UP-SHALE				
D2		TRIAS ARGILO-EVAPORITIC SERIES		KEUPER-MUSHELKALK
TS4				
LOWER SHALE				
T2		TRIAS ARGILO-SANDSTONE SERIES		BUNTSTAND-STEIN
T1				
LOWER SERIES				
PALEOZOIC				

subsidence, were primarily used to define different sedimentary basins which can be grouped into three regions: the **western** (Tindouf, Reggane, Bechar basins); the **northeast central** (Triassic of Hassi R'Mel and Oued M'ya basins); and the **eastern** (Ghadames and Iliizi basins). Actually, the geologic provinces encompass several basins and can be considered as "multi-basin provinces;" on the other hand, some of the basins share with more than one province (Fig. 6.5).

In general, it might be best if we define the geological provinces in relation to the main regions of the Algerian Saharan Platform: the Western, the Northeastern Triassic, and the Eastern regions.

6.2.3.1 The Western Saharan Region

The Western Saharan region mainly consists of the Grand Erg/Ahnet Province that includes the Tindouf, Reggane, and Bechar basins to the west of Algeria, whereas the Triassic basin encompasses the northeastern portion of the Grand Erg/Ahnet province as well as the northwestern portion of the Trias-Ghadames province.

The Grand Erg/Ahnet Province, located primarily in western Algeria, is chiefly a gas producing province but remained practically unexplored. The northwest tip of the province extends slightly into Morocco. The province area encompasses approximately 700,000 km^2, contains the Timmimoun, Ahnet, Sbaa, Mouydir, Benoud, Bechar, Abadla basins and part of the Oued M'ya basin; it also includes the giant Hassi R'Mel gas field.

The Grand Erg/Ahnet province is mainly a Precambrian basement unconformably overlain by thick Paleozoic sediments, structurally defined by four basins (reservoir rocks of Cambro-Ordovician, Silurian, Devonian, and Carboniferous age), ranging from Cambrian to Namurian, that are separated by high zones, followed by insignificant thicknesses of Mesozoic and Cenozoic sediments.

More than one total petroleum system may exist within each of the basins in the Grand Erg/Ahnet Province. The *composite total* petroleum systems are

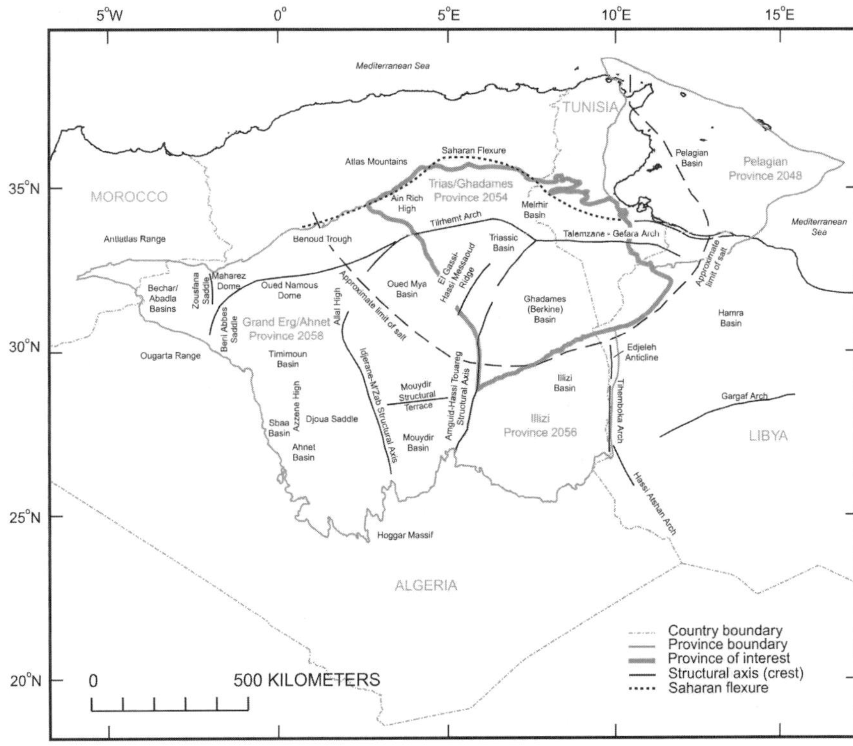

Fig. 6.5 North-central Africa, showing USGS-defined geologic provinces and major structures (modified from Aliev and others, 1971; Burollet and others, 1978; Montgomery, 1994; Petroconsultants, 1996b; Persits and others, 1997)

coincident with the basins in which they exist and are known as Tanezzuft-Timimoun, Tanezzuft- Ahnet, Tanezzuft-Sbaa, Tanezzuft-Mouydir, Tanezzuft-Benoud, Tanezzuft-Bechar-Abadla, and Tanezzuft-Oued M'ya total petroleum systems. The latter extend across both the Grand Erg/Ahnet and the neighboring Trias/Ghadames Province. Tanezzuft refers to the Tanezzuft formation (Silurian), which is the oldest major source rock in the system; in the total petroleum system, Tanezzuft is then followed by the basin name in which the total petroleum system exists. The assessment units are named after the total petroleum systems with a suffix of "structural/stratigraphic," which refers to the progression from a structural and combination trap exploration strategy to a stratigraphic trap exploration strategy.

6.2.3.2 The Northeast Triassic Region

The **Triassic Region of Algeria** divides the Northern Sahara into eastern and western basins where Paleozoic strata form a north-south trending ridge along the longitude of Algiers. The Triassic region is large, an east-west oriented anticlinorium, and is located on the northeast central portion of the Saharan Platform.

The system should be considerably more effective than the traditional methods in performing certain geological tasks, normally spent in nonprofessional work (Assaad, 1981).

6.2.3.3 The Eastern Region

The Eastern Region of the Ghadames-Illizi syncline, which is known as the East Algerian Syncline, consists of both the Illizi and Ghadames basins, and is separated by the Ahara ridge. The Ghadames-Illizi Syncline is bounded on the west by Amguid-El Biod dorsal and by the eastern border of Algeria.

The Trias/Ghadames Province is a geologic province delineated by the USGS located in central east Algeria, southwest of Tunisia and west of Libya. The southern and southwestern boundaries of the Trias/Ghadames province represent the approximate extent of Triassic and Jurassic evaporates that were deposited within the Triassic Basin (Pub.usgs.gov/bul/b2202-c/b2202-bso.pdf).

The province includes the Melrhir Basin, the Ghadames (Berkine) Basin, and part of the Oued M'ya Basins. Although several total petroleum systems may exist within each of these basins, only three "composite" total petroleum systems have been identified. Each total petroleum system occurs in a separate basin, and comprises a single assessment unit. The composite systems are known as (a) the Tanezzuft-Oued M'ya, (b) the Tanezzuft-Melrhir, and (c) the Tanezzuft-Ghadames total petroleum systems; the last two systems are located almost entirely within the Trias/Ghadames Province. The Tanezzuft-Oued M'ya petroleum system extends into the neighboring Grand Erg/Ahnet Province, whereas the Trias/Ghadames Province contains the giant Hassi Messaoud oil field, in the Tanezzuft-Oued M'ya total petroleum system.

The Illizi Province is located southeast of Algeria and a small portion of western Libya. The province and its total petroleum system coincides with the Illizi Basin. The Illizi province contains both the Tin Foyé-Tabenkort and Zarzaitine oil fields. More than one total petroleum system may exist within the Illizi Basin. One "composite" total petroleum system is described as the Tanezzuft-Illizi Total Petroleum System. In the Illizi basin, the Tanezzuft is sometimes referred to as Argillaceous or Argileux. The Tanezzuft Formation is the principal source rock, deposited during a major regional flooding event.

6.2.4 An Approach to a Computerized Lithostratigraphic Chart in North Africa

Geological and geophysical activities greatly increased in northwest African countries over the past three centuries. A precise interpretation of the regional geological framework of the sedimentary formation in different regions of the Saharan Platform of the Arabian Maghreb was carried out, and rock units were better correlated on a regional scale and cited in their proper stratigraphic levels. A regional updated terminology of rocks is available in place of the misleading abbreviations that had been submitted by previous foreign contractors.

A lithostratigraphic chart was built up to define the stratigraphic level of each rock unit among different structural regions of the Algerian Saharan Platform by correlating these rock units among the different regions of the eastern Sahara. These comprise the Oued M'ya-M'Zab region to the northwest; the Erg Oriental-Agreb el-Gassi and Hassi Messaoud region to the southeast; and the Tilrhemt-Illizi region to the south (chart 6.1; charts 6.1a–c).

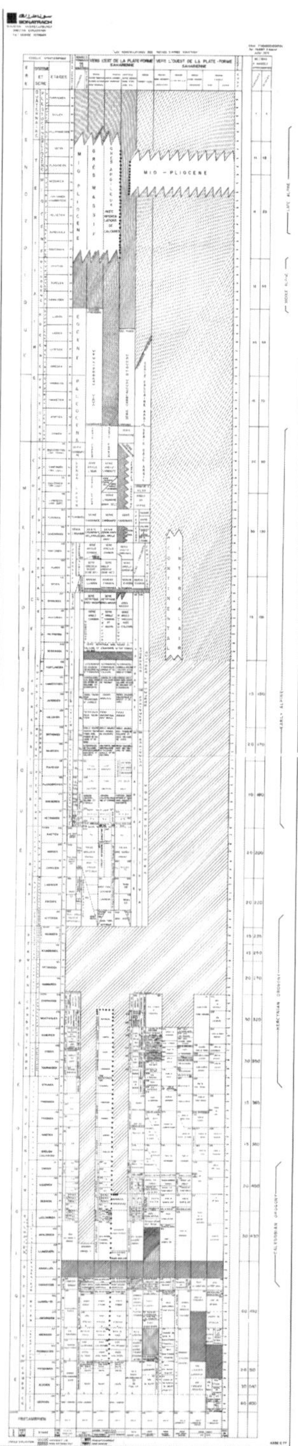

Chart 6.1 A lithostratigraphic correlation chart of Sedimentary Section – North east centre of Algerian Sahara; (**a**) a lithostratigraphic correlation of the Cenozoic Sedimentary Section of the Algerian Saharan Platform; (**b**) a lithostratigraphic columnar Section of the Mesozoic "Algerian Saharan Platform"; (**c**) a lithostratigraphic columnar Section of the Paleozoic "Algerian Saharan Platform"

Chart 6.1a

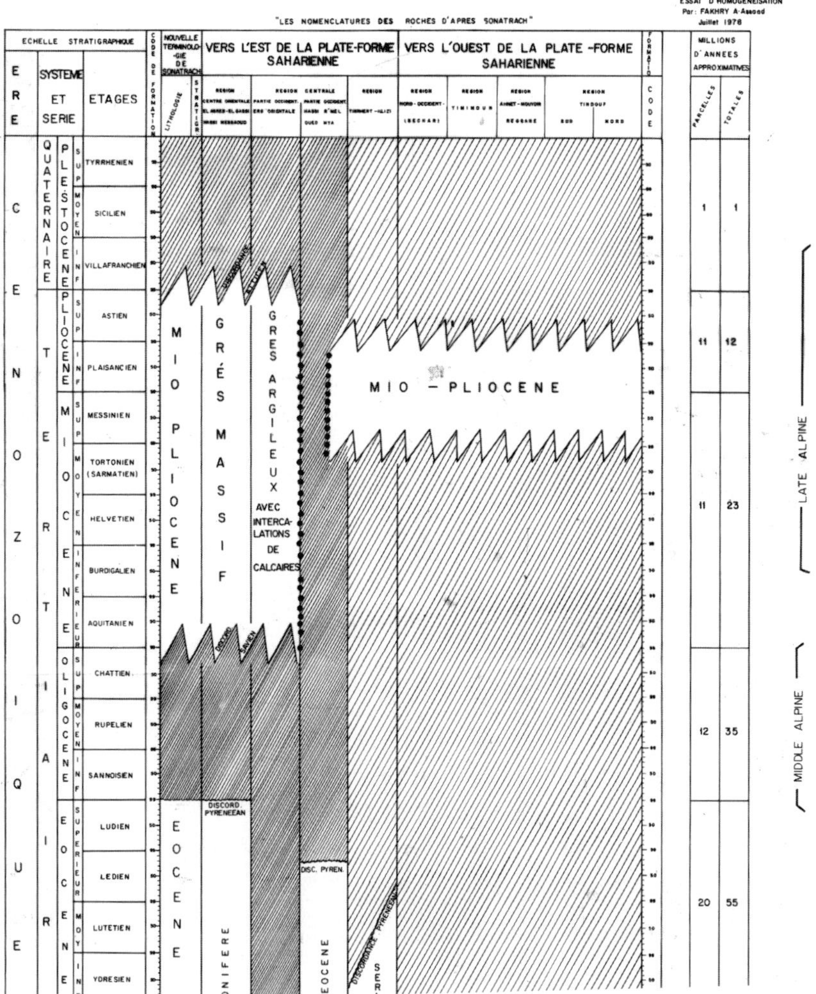

6.2.4.1 Numeric Coding of Different Elements of the Geologic Provinces of the Algerian Sahara

Klett (2000) designed a numeric code to identify each region, province, total petroleum system, and assessment unit. A graphical depiction that places the elements of the total petroleum system into the context of geological time is provided in the form of an events chart, the items of which include the major rock unit names; the temporal extent of source rock deposition, reservoir rock deposition, seal rock deposition, and overburden rock deposition; and the generation, migration, accumulation, and preservation of petroleum, defined in a petroleum system (Magoon and Dow 1994).

A numeric code is given to identify each region, province, total petroleum system, and assessment unit of the Erg/Ahnet province, Illizi Province, and Ghadames Province.

Table 6.1 A numeric code applied for the Grand Erg/Ahnet Province, Algeria and Morocco, and its total petroleum systems [pubs.usgs.gov/bul/b2202-b/b2202-bso.pdf]

Table 6.2 A numeric code applied for the Illizi province [pubs.usgs.gov/bul/b2202-a/b2202-bso.pdf]

Chart 6.1b

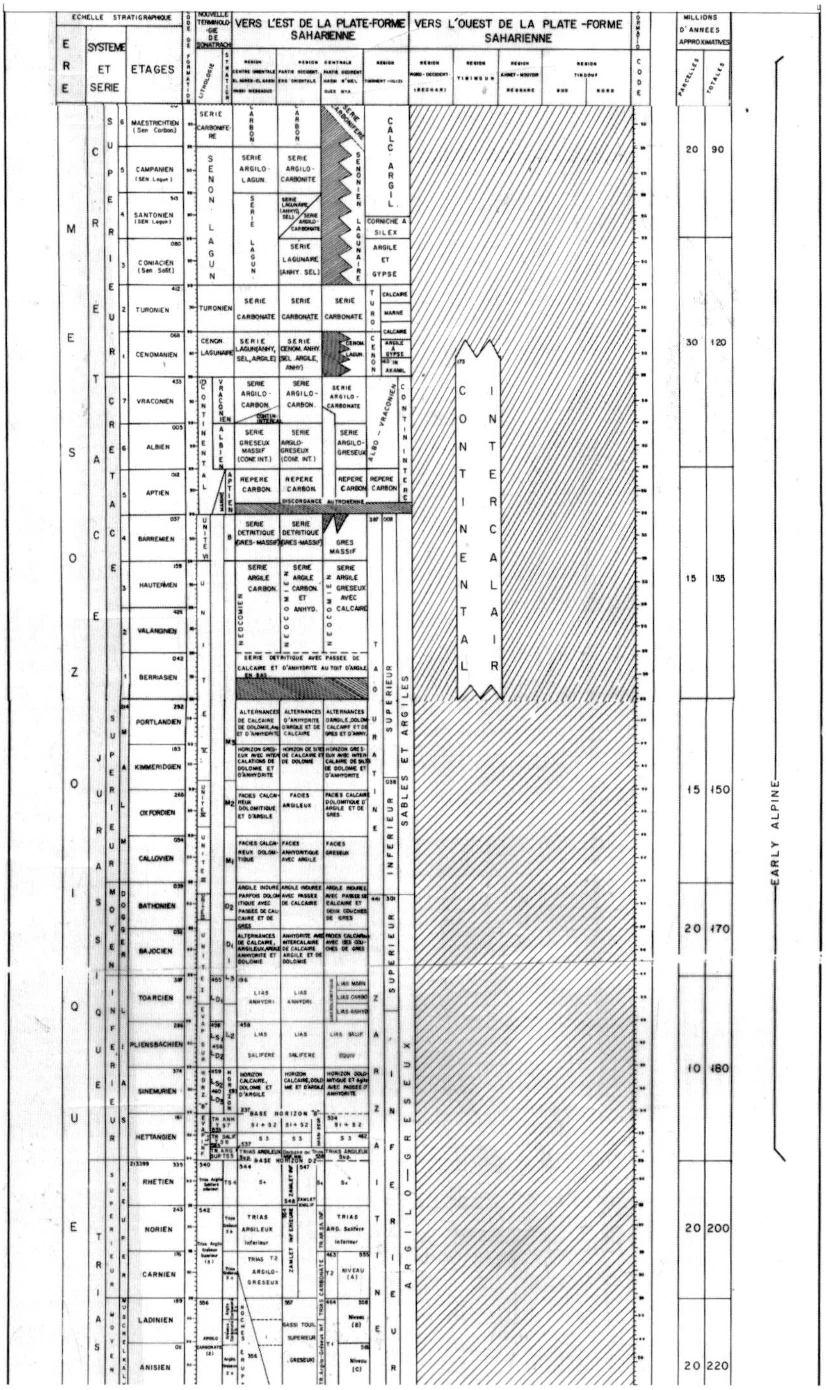

ERE	SYSTEME ET SERIE	ETAGES		VERS L'EST DE LA PLATE-FORME SAHARIENNE			VERS L'OUEST DE LA PLATE-FORME SAHARIENNE				CODE	MILLIONS D'ANNEES APPROXIMATIVES	
				RÉGION CENTRE-ORIENTALE	RÉGION CENTRALE	RÉGION	RÉGION NORD-OCCIDENT	RÉGION TINRHERT	RÉGION AHNET-MOUYDIR	RÉGION TINDOUF		PARCELLES	TOTALES
M E S O Z O I Q U E	C R E T A C E	S U P E R I E U R	MAESTRICHTIEN (Sen Carbon.)	SÉRIE CARBONIFÈRE			CALC. ARGIL.					20	90
			CAMPANIEN (Sen Lagun.)	SÉRIE ARGILO-LAGUN.	SÉRIE ARGILO-CARBONITE								
			SANTONIEN (Sen Argilo)		SÉRIE LAGUNAIRE (ANHY.) SÉRIE ARGILO-CARBONATE		CORNICHE A SILEX						
			CONIACIEN (Sen Salif.)		SÉRIE LAGUNAIRE (ANHY. SEL.)		ARGILE ET GYPSE						
			TURONIEN	SÉRIE CARBONATE	SÉRIE CARBONATE	SÉRIE CARBONATE	CALCAIRE MARNE CALCAIRE					30	120
			CENOMANIEN	SÉRIE LAGUN.(ANHY. SEL., ARGILE)	SÉRIE CENOM. ANHY. (SEL. ARGILE, ANHY.)		ARGILE & GYPSE						
			VRACONIEN	SÉRIE ARGILO-CARBON.	SÉRIE ARGILO-CARBON.	SÉRIE ARGILO-CARBONATE							
			ALBIEN	SÉRIE GRESEUX MASSIF (CONE INT.)	SÉRIE ARGILO-GRESEUX (CONE INT.)	SÉRIE ARGILO-GRESEUX						15	135
			APTIEN	REPÈRE CARBON.	REPÈRE CARBON.	REPÈRE CARBON.	REPÈRE CARBON.						
		I N F E R I E U R	BARREMIEN	SÉRIE DETRITIQUE (GRES-MASSIF)	SÉRIE DETRITIQUE (GRES-MASSIF)	GRES MASSIF							
			HAUTERIVIEN	SÉRIE ARGILE CARBON.	SÉRIE ARGILE ET ANHYD.	SÉRIE ARGILE GRESEUX AVEC CALCAIRE						15	135
			VALANGINIEN										
			BERRIASIEN	SÉRIE DETRITIQUE AVEC PASSÉE DE CALCAIRE ET D'ANHYDRITE AU TOIT D'ARGILE									
	J U R A S S I Q U E	S U P E R I E U R	PORTLANDIEN	ALTERNANCES DE CALCAIRE DE DOLOMIE ET D'ANHYDRITE	ALTERNANCES D'ANHYDRITE D'ARGILE ET DE CALCAIRE	ALTERNANCES D'ARGILE, DOLOMIE CALCAIRE ET DE GRES ET D'ANHY.						15	150
			KIMMERIDGIEN	HORIZON GRESEUX AVEC INTERCALATIONS DE DOLOMIE ET D'ANHYDRITE	HORIZON DE GRES AVEC DOLOMIE	HORIZON GRESEUX AVEC INTERCALAIRE DE DOLOMIE ET D'ANHYDRITE							
			OXFORDIEN	FACIES CALCAIREUX DOLOMITIQUE ET D'ARGILE	FACIES ARGILEUX	FACIES CALCAIRE DOLOMITIQUE D'ARGILE ET DE GRES							
		M O Y E N	CALLOVIEN	FACIES CALCAIREUX DOLOMITIQUE	FACIES ANHYDRITIQUE AVEC ARGILE	FACIES GRESEUX							
			BATHONIEN	ARGILE INDURÉE PARFOIS DOLOMITIQUE AVEC PASSÉE DE CALCAIRE ET DE GRES	ARGILE INDURÉE AVEC PASSÉE DE CALCAIRE	ARGILE INDURÉE AVEC PASSÉE CALCAIRE ET DEUX COUCHES DE GRES						20	170
			BAJOCIEN	ALTERNANCES DE CALCAIRE, ARGILEUX, ANHYDRITE ET DOLOMIE	ANHYDRITE INTERCALAIRE ARGILE ET DE DOLOMIE	SÉRIES CALCAIRE AVEC DEUX COUCHES DE GRES							
		I N F E R I E U R	TOARCIEN	LIAS ANHYDRI	LIAS ANHYDRI	LIAS MARN LIAS CARBO LIAS ASHYD							
			PLIENSBACHIEN	LIAS SALIFERE	LIAS SALIFERE	LIAS SALIF EQUIV						10	180
			SINEMURIEN	HORIZON CALCAIRE, DOLOMIE ET D'ARGILE	HORIZON CALCAIRE, DOLOMIE ET ANHY ET D'ARGILE	HORIZON DOLOMITIQUE ET ARGILE AVEC PASSÉE D'ANHYDRITE							
			HETTANGIEN										
T R I A S		S U P E R I E U R	RHETIEN									20	200
			NORIEN	TRIAS ARGILEUX Inférieur		TRIAS ARG. Salifère Inférieur							
			CARNIEN	TRIAS T2 ARGILO-GRESEUX		NIVEAU (A)							
		M O Y E N	LADINIEN	ARGILO CARBONATE		GASSI TOUIL SUPERIEUR Niveau (B)						20	220
			ANISIEN			GRESEUX Niveau (C)							

Chart 6.1c

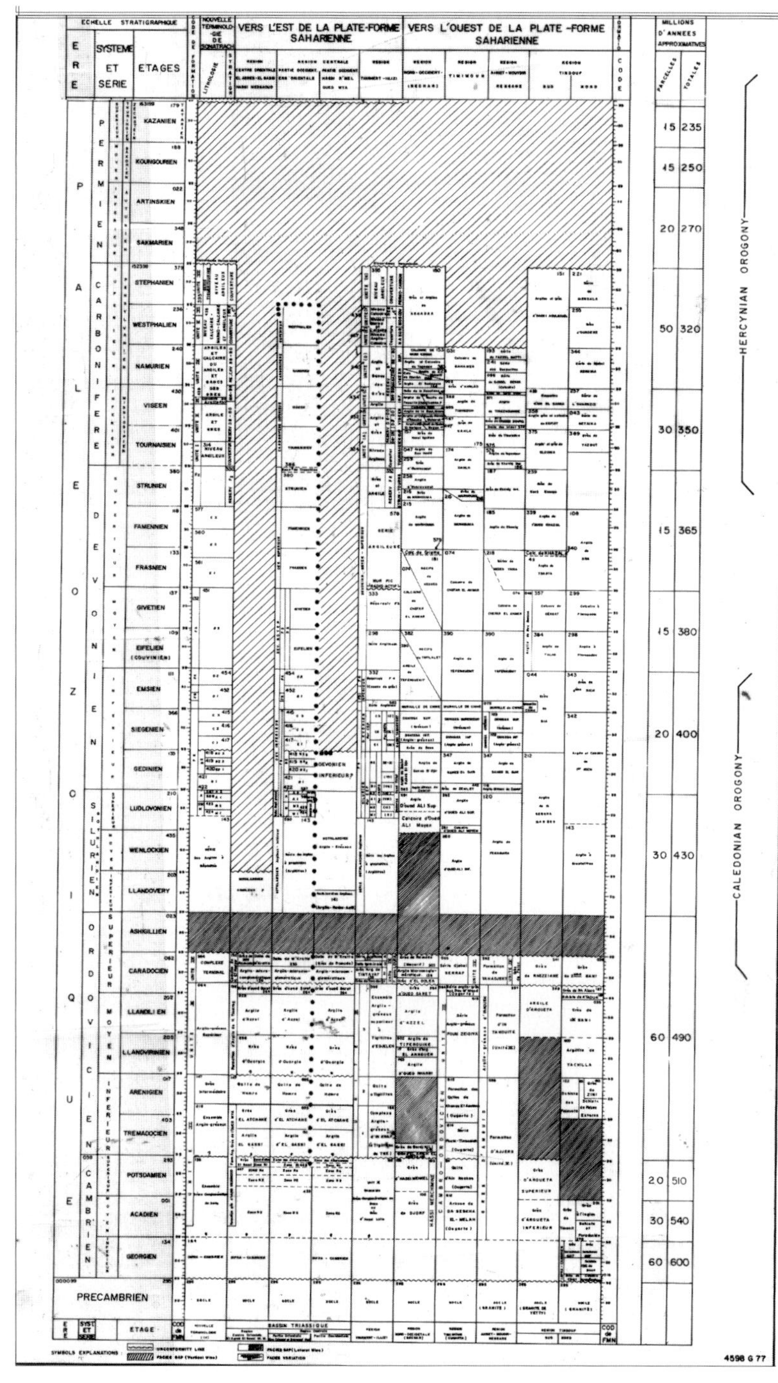

Table 6.1 A numeric code applied for the Grand Erg/Ahnet Province, Algeria, and Morocco and their total petroleum systems (Klett 2000)

Unit	Name	Code
Region	Middle East and North Africa	2
Province	Trias/Ghadames	2058
Total petroleum systems	a- Tanezzuft-Timimoun	205801
	b-Tanezzuft-Ahnet	205802
	c-Tanezzuft-Sabaa	205803
	d- Tanezzuft-Mouydir	205804
	e- Tanezzuft-Benoud	205805
	f – Tanezzuft-Bechar/Abadla	205806
Assessment units	a- Tanezzuft-Timimoun Struc/Stratig	20580101
	b- Tanezzuft-Melrhir Struc/Stratig	20580201
	c- Tanezzuft-Ahnet Struc/Stratig	20580301
	d. Tanezzuft-Mouydir Struc/Stratig	20580401
	e- Tanezzuft-Benoud Struc/Stratig	20580501
	f- Tanezzuft-Benoud Struc/Stratig	20580601

Table 6.2 A numeric code for the region, province, total petroleum systems, and assessment units of the Illizi basin (Klett 2000)

Unit	Name	Code
Region	Middle East and North Africa	2
Province	Illizi	2056
Total petroleum system	Tanezzuft-Illizi	205601
Assessment unit	Tanezzuft-Illizi Structural/Stratigraphic	20560101

Table 6.3 A numeric code for the region, province, total petroleum systems, and assessment units of the Trias/Ghadames Province (Klett 2000)

Unit	Name	Code
Region	Middle East and North Africa	2
Province	Trias/Ghadames	2054
Total petroleum systems	a- Tanezzuft-Oued M'ya	205401
	b-Tanezzuft-Melrhir	205402
	c-Tanezzuft-Ghadames	205403
Assessment units	Tanezzuft-Oued M'ya Struct/Stratig	20540101
	b-Tanezzuft-Melrhir Struct/Stratig	20540201
	c-Tanezzuft-Ghadames Struct/Stratig	20540301

Table 6.3 A numeric code applied for the Trias/Ghadames province in Algeria, Tunisia, and Libya, including the Tanezzuft-Oued M'ya, Tanezzuft-Melhir, and Tanezzuft-Ghadames [pubs.usgs.gov/bul/b2202-c/b2202-bso.pdf]

The criteria for assigning codes are uniform throughout all publications of the USGS-WPA 2000. The author considers such numeric codes useful for further application of the proposed computerized lithostratigraphic correlation chart of the Arabian Maghreb in northwestern Africa, to be extended to cover North Africa as well.

6.2.4.2 General Structural Settings of the Algerian Sahara

The primary sedimentation of the Saharan platform has been disturbed by deep-seated movements that separated the platform into more or less well defined basins, each with its own specific geologic history and each posing different problems for petroleum exploration.

The Algerian Saharan platform is generally classified into three major stages; the Precambrian, the Paleozoic, and the Mesozoic. The **Precambrian formation** constitutes the basement rock, and is formed of strongly dislocated metamorphosed and granitized rocks outcropping at the Hoggar massif to the far southeast of Algeria. The **Paleozoic** and **Mesozoic formations** include the main reservoir rocks in the Algerian Saharan platform and can be classified into three types of structures: (1) High structures comprise Hassi Er R'Mel Mole, Hassi Messaoud, Tilrhemt, Rhoured El- Begoul, Tinfoye', etc. (2) Low structures comprise Bechar, Tindouf, Reggane, Oued M'ya, and Rhadames Basins. (3) Structural terraces include the central structure of Illizi, and the structural terrace of Mouydir (Fig. 6.6). The Paleozoic-Mesozoic reservoir rocks are oil and gas productive in three main regions: the Triassic Province of Oued M'ya-M'zab and Hassi Messaoud to the northwest, the Illizi basin to the southeast, and the Paleozoic basins in isolated regions to the southwest (Tindouf, Bechar, Regane, and Timmimoune).

6.2.4.2.1 The Triassic Region of Oued M'ya-M'Zab

The Triassic Province was so named because of thick Triassic-Liassic evaporite deposits that overlie the Triassic detritals. It comprises three main structures: (1) the Hassi R'Mel Domal structure to the northwest, and the Oued M'ya–M'Zab structure in the central portion, where the Oued Noumer, Ait Kheir, and Djorf oil and gas fields are located; (b) the Erg-Oriental and

El-Borma oil fields capped by Triassic limestone, located to the southeast; and (c) the Haoud Berkoui and Haniet El-Mokta oil fields to the east of Oued M'ya, producing from the Lower Triassic sandstone series (SI-series), and capped by eruptive rocks.

6.2.4.2.2 The Hassi Messaoud–El Agreb–El Gassi

The Hassi Messaoud–El Agreb–El Gassi to the southeast of Hassi R'Mel, consists of different fields where the massif reservoir of the Cambrian formation is mainly composed of fissured quartzites. Oil accumulation in the Hassi Messaoud structure results from a combination trap where the crest is associated with the Hercynian unconformity. The Cambrian reservoirs ($R_1 + R_2$), are overlain by Late Triassic-Liassic deposits.

6.2.5 Stratigraphy of the Algerian Saharan Platform

A brief summary of different stratigraphic stages, systems, and series is given in chronological sequence (**Chart 6.1** is a much reduced chart of the whole sedimentary section (CD-ROM Filename: Stratigraphy.)

6. The **Aschgillian** stage of the upper Cambro-Ordovician formations is not represented; the

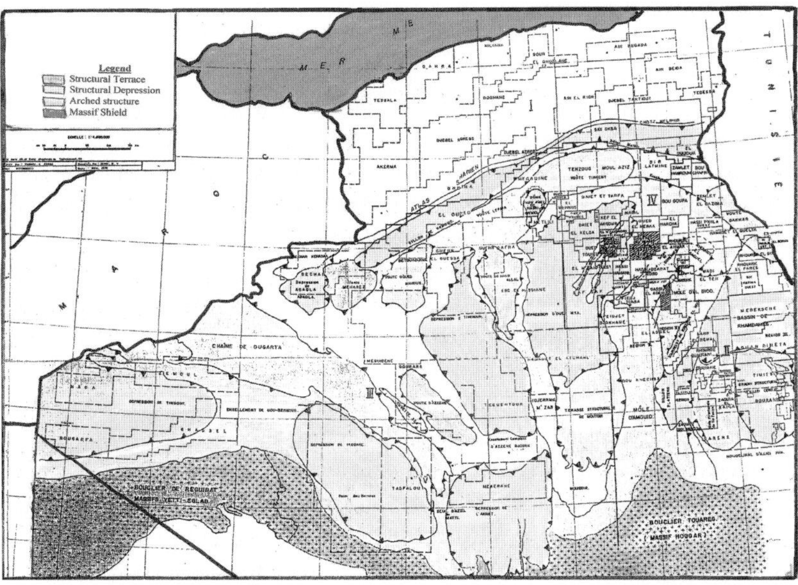

Fig. 6.6 Regional structural map of the Algerian Sahara

uppermost rock units are related to the middle and upper Caradocian.

5. The **Carboniferous** section at Illizi Basin (Lower Tournasian stage to Upper Stephanian stage) was classified according to palynological studies; the petroleum reservoirs D6–D8 series are related to the Upper Tournasian stage. At Tindouf basin, the Upper and Lower Visean reservoirs attained their maximum thickness; whereas the upper Numerian and Westphalian formations are not represented in some locations (Chart 6.1c).

4. The **Buntsandstein** stage is related to the reworked detrital series of Lower Triassic (Scythenian) or Lower series "SI." The Mushelkalk stage belongs to the Shaley Carbonate Triassic formations or its equivalent "T1- unit." The lowermost stage of Keuper, "Carnian," is related to the upper detrital Triassic "T2-unit." The uppermost stage of Keuper, "Rhetian," is related to the massive salt bed "TS4" which is underlain by the marker bed "D2," on top of the Triassic formation.

Recent studies considered the whole Triassic section as Keuper in age "marker bed of D2." The Lower Liassic section "S1+S2+S3+upper shales" is equivalent to the Hettangian Stage, according to palynological and regional stratigraphic studies; the top of Hettangian is at the base of another marker bed: "Horizon B." At Illizi Basin, the Triassic and Lower Liassic sections are related to the Lower Zarzaitene, whereas Upper Liassic and Dogger formations represent the Upper Zarzaitine formation (see Fig. 6.4; Assaad 1972).

3. **Upper Toarcian—Top Barremian (Upper Liassic to Base Aptian)** formations are classified according to regional electric log correlation into six rock units (Units I–VI). The lower evaporitic Liassic section **"Pliensbachian"** is considered the Upper Evaporite unit. Each of these rock units of the Jurassic-Lower Cretaceous formations shows a complete cycle of sedimentation to the north where the transgression and regression from and to the north occurred, and ended with sea regressions where continental deposits then prevailed.

2. **The Middle and Upper Cretaceous formations "Aptian-Maestrichtean"** are studied to define stratigraphic levels of different rock units, regional structural trends, the evolution and thickness of facies, the environment of deposition and paleo-geographic evolution. The carbonate barrier base of the Aptian formation, overlying the Austrichtean discordance, separates the continental sandstone masses of the Albian and Barremian formations (**Chart 6.1b**).

1. **The Miopliocene** eroded surface (Aquitanian to Astian) outcrops in some portions of the Algerian Sahara (**Chart 6.1a**).

6.2.6 A Geological Study of An Exploratory Well at the Eastern Border of the Algerian Sahara (Ry-1)*

The Rhourde Yacoub petroleum exploration well (Ry-1) was completed in February 1972 at the eastern border of the Algerian Saharan platform that extends to the Tunisian Sahara. The Miopliocene outcrops at the location of the well, and was completed in the Lower Devonian at a total depth of 4,340 m. The Triassic sandstone reservoir yields salt water, whereas the Middle Devonian produces poor gas and gazoline; the Lower Devonian yields saltwater. Table 6.4 shows the general well data of the exploratory well Ry-1. Figures 6.7 and 6.8 show the location map of the Ry-1 well and a cross section among nearby wells, respectively; both figures were submitted by the Directorate of Studies and Synthese of Sonatrach (Chebourou[*] H, personal communication).

6.3 Petroleum Geology of the Libyan Sahara

6.3.1 Scope

North Africa contains two outstanding provinces: the Sirte basin of Libya and the Triassic/Illizi basins of Algeria; both the Sirte and Triassic/Illizi provinces contain reserves in the range of 30–35 BBOE, placing them amongst the world's largest; together they contain 85% of the oil and 80% of the gas discovered in North Africa. The African Sahara contains the largest oil and gas fields in the North African region: the Hassi Messaoud and Hassi R'Mel, which were producing

---

[*] Chebourou, H.-Director of Research and Studies, Sonatrach, Boumerdas, Alger

Table 6.4 General well data of the well Rhourde Yacoub (Ry-1)

Formation tops	Thickness	From	To	Remarks
Mio-Pliocene	(30 m)	00 (279)	35 (+244) m	Friable Sandstone
Senonian Carbonate	(333 m)	35 (244)	368 (−89) m	Calcareous
Senonian Lagunaire	(153 m)	368 (−89)	521 (−242) m	Salt & Anhydrites
Turonian	(89 m)	521 (−242)	610 (−331) m	Limestone
Cenomanian	(174 m)	610 (−331)	784 (−505) m	Brown Shaley ss
Varconian	(26 m)	784 (−505)	810 (−351) m	Dolo. Brown clay
Albian	(95 m)	810 (−351)	905 (−626) m	Sandstone and silt
Aptian	(30 m)	905 (−626)	935 (−656) m	Compact dolomite
Barremian	(141 m)	935 (−656)	1,076 (−797) m	Shaley sandstone
Neocomian	(219 m)	1,076 (−797)	1,295 (−1,096) m	Silty shale and ss
Malm	(197 m)	1,295 (−1,096)	1,492 (−1,213) m	Silt, shale and ss
Dogger Argileux	(74 m)	1,492 (−1,213)	1,566 (−1,287) m	Shale and silt
Dogger Lagunaire	(186 m)	1,566 (−1,287)	1,752 (−1,473) m	Anhydritic Sh. & silt
Lias Anhydrite	(135 m)	1,752 (−1,473)	1,887 (−1,608) m	Anhydrite and shale
Lias Salifere	(49 m)	1,887 (−1,608)	1,936 (−1,657) m	Shaley salt
Horizon (B)	(20 m)	1,936 (−1,657)	1,956 (−1,677) m	Limestone marker bed
Trias evaporites	(115 m)	1,956 (−1,677)	2,071 (−1,792) m	Shaley anhydrite & salt
Trias Carbonate	(66 m)	2,071 (−1,792)	2,137 (−1,858) m	Shaley carbonate
Lower Trs. Sh. ss	(103 m)	2,137 (−1,858)	2,240 (−1,961) m	Sandstone, shaley
Lower Carbonifere	(679 m)	2,240 (−1,961)	2,919 (−2,640) m	Ss, silt & shale
Up. Devonian	(604 m)	2,919 (−2,640)	3,523 (−3,244) m	Sandy Shale
Middle Devonian	(468 m)	3,523 (−3,244)	3,991 (−3,712) m	Calcareous shaley ss
Lower Devonian	(>394 m)	3,991 (−3,712)	TD = 4,340 m	sandy shale

Ground surface (Zs) = 274.34 m
Rotary table (Zt) = 279.08 m
Coordinates: X = 9° 20′ 33″E
Y = 30° 46′ 08″N

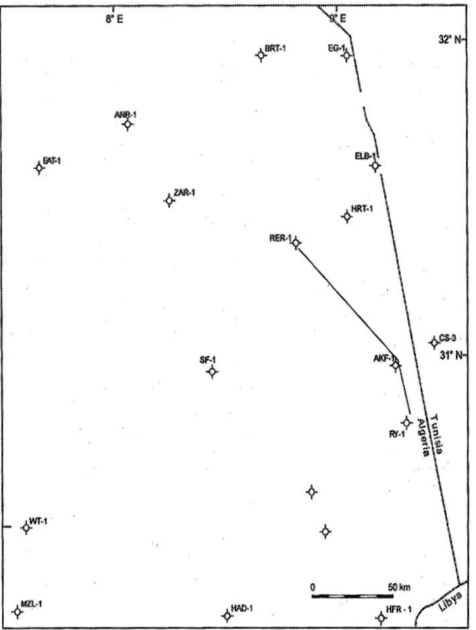

Fig. 6.7 Location map of well Ry-1 and cross section

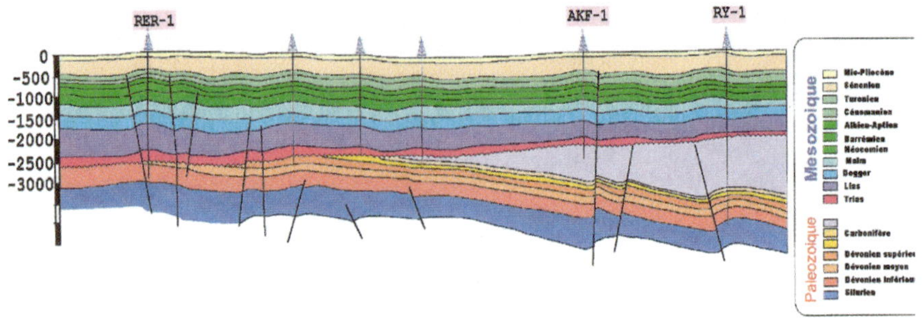

Fig. 6.8 Cross section in Berkine basin (After N. Goucem, Sonatrach, 2000)

approximately 9000 BBO and 60 TCFG, respectively. The Gulf of Suez, though significantly smaller in terms of total reserves, is considered the third most significant province.

6.3.2 Stratigraphy of Sebha Area West of Libya*

6.3.2.1 Discussion

The Regwa Company, of Cairo, Egypt, carried out a stratigraphical study of the Sebha area. The surface geology of the area, located between latitudes 25°–27°30′N and longitudes 12°30′–15°E, includes three main Wadis—El-Ajal, Shatti, and Traghen (Werwer* 1973; Fig. 6.9).

The **outcropping formations** range from Cambrian to Recent and are generally interrupted by unconformities:

1. Quaternary deposits (Early Tertiary)
2. Murzuk formation (Lower Cretaceous)
3. Nubian sandstone (undeveloped because of unconformity)

---

* Geologist Ali Werwer, Chairman and CEO (REGWA, 1973), Cairo, Egypt.

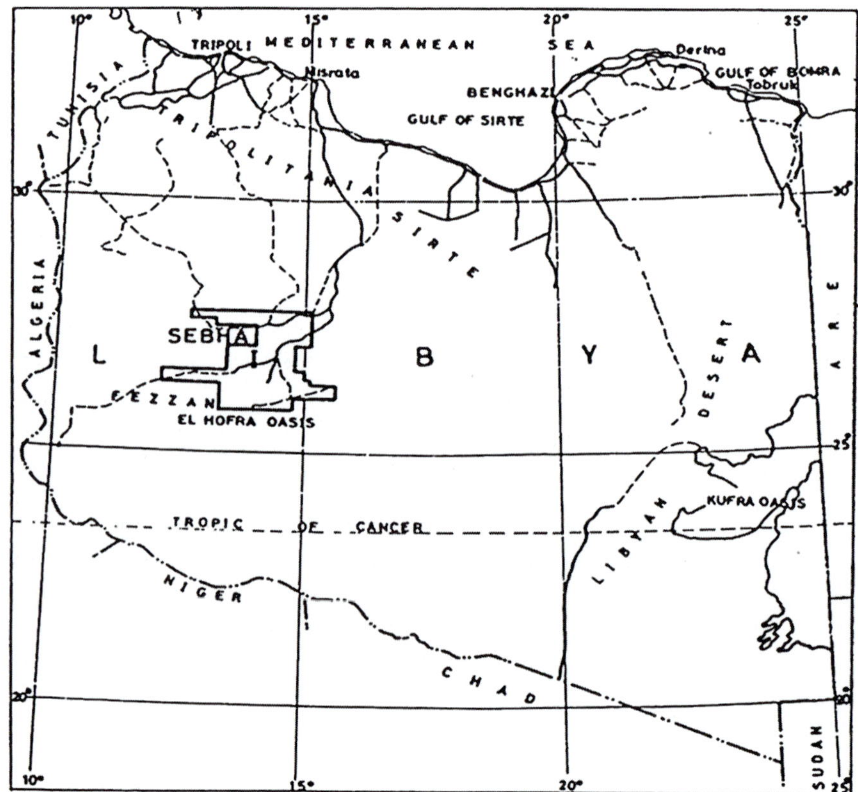

Fig. 6.9 Location map of Sebha Area, Libya

4. M'rar formation-Uwaynat Wannin formation (Upper Devonian to Lower Carboniferous)
5. Memouniat formation (Upper Devonian)
6. Haouaz formation (undeveloped because of unconformity)
7. Hassouna formation (Upper Ordovician)
8. Cambrian-Lower Middle Ordovician rocks (unidentified) and overlain by unconformity.

A **detailed lithologic description** of formations was prepared from drilling 39 test wells; vertical and lateral facies changes of rock units were given:

1. Murzuk formation
   ~ unconformity
2. Nubian sandstone
   ~ unconformity
3. Zarzaitine post-Tassilian group
   (b) El-Regeba green clay member
4. Continental post-Tassilian group
5. Tournaisian post-Tassilian group
   ~ unconformity
6. Tiguentourine post-Tassilian group
   (a) Temenhent red clay member
7. Dembaba formation (Upper Carboniferous)
8. Assedi Jefar formation-M'rar formation (Lower Carboniferous)
9. Uwaynat Wennin formation (Middle Upper Devonian)
10. Cambro-Ordovician formation

Table 6.5 Results of micropalaentological study in some wells of Sebha area

Table 6.5 Formation tops from micropalaeontogical results in some wells at Sabha area

Well No.	Depth	Age
12	600–606	Lower Cretaceous
12	669–672	Lower Jurassic
12	686–692	Jurassic
9	400–406	Jurassic
9	501–507	Lower Carboniferous
23	218–224	Upper lower Carboniferous
23	308–314	
23	358–364	Upper lower Carboniferous
23	403–409	
		Upper lower Carboniferous
		Upper lower Carboniferous

6.3.2.2 A General Lithostratigraphic Comparison of the Sedimentary Section West of the Libyan and Northeast of the Algerian Sahara

The sedimentary section of the Libyan Sahara can be summarized in chronological sequence:

10. **Caradocian of Upper Ordovician**: The Hassouna formation of Libya can be compared with Caradocian of the Upper Ordovician in the Saharan platform.
9. **Frasinian-Givitian of Middle-Upper Devonian**: The Awenat Wennin formation (Middle-Upper Devonian) of the Libyan territory can be compared with the Frasinian-Givitian of the Saharan platform.
8. **Strunian-Tournasian**: The Mrar and Assed Jefar formation of the Libyan territory can be compared with the Upper Devonian-Lower Carboniferous formations (Strunian-Tournasian) of the Algerian Saharan platform.

   The Carboniferous section at the Illizi basin of the Algerian Sahara was classified according to palynological studies as Upper Tournasian. At Tindouf basin, the Upper and Lower Visean reservoirs attained their maximum thickness, whereas the Upper Numerian and Westphanian formations are not represented in some locations.
7. **Stephanian of the Upper Carboniferous**: The Dembaba formation of the Libyan territory can be compared with the Upper Carboniferous (Stephanian) of the Algerian Saharan platform, whereas the Tiguintourine formation is represented in both territories as Upper Carboniferous and is differentiated in the Libyan Territory by two clay members: a lower El-Regeba green clay overlain by Temenhert red clay—"compare with the variegated shales (Maestrichtian of the Upper Cretaceous?) overlying the Nubian sandstone aquifer of the Kharga oases of the Egyptian Desert." (Assaad, 1988)
6. **The LowerTriassic Shaley Sandstone** (Scythian)—equivalent to "SI" series of the Algerian Saharan platform as reworked detrital series of the Buntsandstein stage; the Mushelkalk stage followed upward by the shaley carbonate Triassic formation or its equivalent "T1-unit," which is overlain by the lowermost stage of

Keuper (Carnian), equivalent to the upper detrital Triassic "T2-unit" series. The uppermost stage of Keuper (Rhetian) is related to the massive salt bed "TS4" that directly underlies the marker dolometic bed "D2," which defines the top of the Triassic formation of the Saharan Platform, according to an updated correlation (Stoica and Assaad 1981).

5. **Liassic-Middle Jurassic**: The Zarzaitine formation of the Libyan territory can be compared with the Lias-Upper Jurassic formation of the Algerian Saharan platform. The lower Liassic section "TS1 + TS2 + TS3 + upper shales" of the Algerian Saharan platform is equivalent to the Hettangian stage according to Palyanological bed "Horizon B." At the Illizi basin, the Triassic and Liassic sections are related to the Zarzaitine formation.

4. **From Top Jurassic to Lower Cretaceous**: This section is only known in the Algerian Saharan platform and is differentiated into six rock units, "Units I–VI," and is classified according to electric log correlation where the underlying evaporites (Pliensbachian stage), are considered the Upper evaporite unit. To the north, each of these rock units of the Jurassic-Lower Cretaceous formations show a complete cycle of sedimentation. The Jurassic-Lower Cretaceous formation ended with sea regression toward the north where continental deposits prevailed.

3. **Lower Cretaceous**: In the Libyan territory, Touratine and Marzouk formations, separated by successive unconformities, can be compared with the Lower Cretaceous of the Algerian Saharan platform.

2. **Middle Cretaceous**: The lowermost carbonate bed of the Aptian formation that overlies the Austrichtian discordance separates the continental sandstone masses of the Albian and Barremian formations of the Middle Cretaceous and can be compared with the continental post-Tassillian group of the Libyan territory. Aptian and Albian formations are good aquifers in the Algerian Saharan platform.

1. **Mio-Pliocene Outcrops**: The eroded surface of the Mio-pliocene outcrops in some parts of the Algerian Sahara. (**N.B**: A precise stratigraphic comparison is highly recommended).

6.4 Petroleum Geology of Egypt, Northeast Africa

There are three main areas of petroleum interest in Egypt: the northern part of the Western Desert, The Nile Delta, and the Gulf of Suez province.

6.4.1 The Western Desert of Egypt*

The geology of the Western Desert of Egypt is greatly different from that of the Western Arabian Maghreb region. The Western Desert of Egypt includes the most fertile desert land and embraces an area of $681,000 \text{ km}^2$, or more than two-thirds of the Egyptian territory. It is one of the most arid regions in the world; the sources of water are found hundreds of km apart. The Western Desert is also one of the hottest places in the world; it is essentially a plateau with extensive areas of rubbly rock surface, covered in some places by long swaths of sand dunes.

Active oil exploration in the vast area of the Western Desert was started in the mid 1950s by the Sahara Petroleum Company (SAPETCO); in contrast, the first well was drilled in the Gulf of Suez in 1908.

The detailed subsurface structural synthesis in the northern part of the Western Desert of Egypt has become of great interest for its petroleum potential, because it comprises the second most promising locality in Egypt after the Gulf of Suez province; three petroleum fields are found:

Two oil and gas fields are under study in El-Razzak and the Abu Gharadig area, and one oilfield in the Meleiha area; the two petroleum areas are located between latitudes $29°30'–30°40'\text{N}$ and longitudes $28°22'–28°38'\text{E}$, covering about 180 km^2 and 240 km^2, respectively; the stratigraphic section in the northwestern desert of Egypt ranges in age from the Cambro-Ordovician to Recent. The Quaternary, Pliocene, and Miocene sediments are exposed on the surface and form the blanket over the subsurface section that has been penetrated by 14 wells in the El-Razzak field area and 18 wells in the Abu-El-Gharadig field area (Fig. 6.10). Most of the wells drilled in the area were completed in the Lower Cretaceous deposits. The reservoir rocks in both fields,

---

* M. Hamed Metwalli, George Philip, A.M.A Wali (1979)

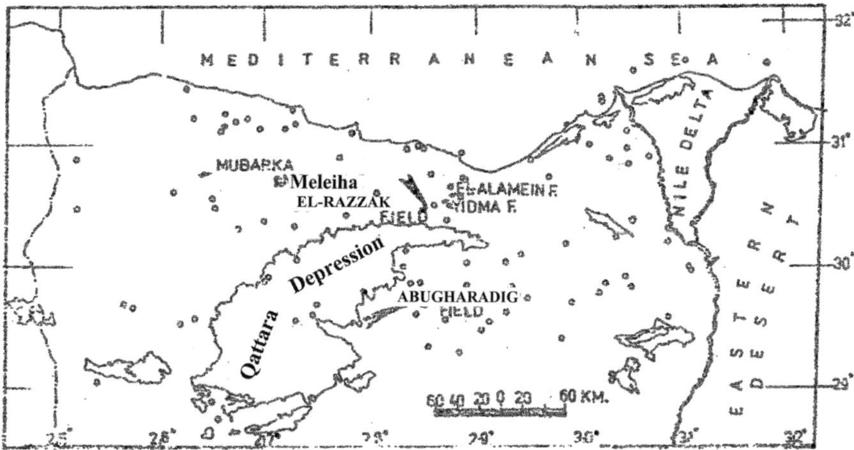

Fig. 6.10 Location of El-Razak and Abu Gharadig Oil and gas fields, northwest of the western desert of Egypt (Metwalli et al. 1979)

of Aptian, Cenomenian, and Turonian clastics and carbonates, generally include a relatively marine and nonmarine series of sedimentary rocks that overlie the basement rocks, and vary in thickness from one locality to another. A detailed study of the structural modeling of the Abu-Gharadig and El-Razzak oil and gas fields has shown that both structures have changed in shape, size, or amplitude and shifted their position laterally as a result of repeated folding stages that affected the area during the Cretaceous age.

A third area, at the Meleiha oil fields, produces oil from the Upper Bahariya formation (Upper Cretaceous/Lower Cenomanian). The field area covers $700\,\mathrm{km}^2$, at latitude $30°36'–30°54'\mathrm{N}$ and longitude $27°00'–27°18'\mathrm{E}$. The Lower Cenomanian (the upper part of the Bahariya formation) is the main producing zone in the Meleiha oil fields. The study of lithofacies, structural features, and electric logging indicates that the sediments were deposited in tidal-flat, near-shore, shallow marine environments.

The Qattara-Siwa Depression occupies about $19,000\,\mathrm{km}^2$ in the northwestern desert, hewed into Miocene rocks. The Qattara depression is the deepest and largest, where the floor descends at its lowest point to $134\,\mathrm{m}$ below sea level, and is very suitable for establishing an electric power system from a waterfall that could be generated from the Mediterranean Sea.

To the north of Qattara rises a third plateau, averaging $200\,\mathrm{m}$ above sea level, which slopes northward toward the coast of the Mediterranean between Alexandria and Sallum. The Qattara Depression contains only salt lakes and salt marshes and is consequently uninhabitable.

The main depressions of the Western Desert are: Kharga-Dakhla, the Bahariya-Farafra oases, the Quattara-Siwa, and the Fayium and Wadi El Natrun depressions; each has its own tradition and dialect.

The total area of the oases is $75,165\,\mathrm{km}^2$; they extend between latitudes $24°30'–30°00'\mathrm{N}$ and longitudes $28°30'–30°35'\mathrm{E}$.

There is a high plateau of Nubian sandstone in the south, extending from the mountains of Uwaynat (over $1,800\,\mathrm{m}$), which descends slowly till it reaches the depression of both the Kharga and Dakhla oases. To the north of the depression, a limestone plateau ($500\,\mathrm{m}$ above sea level) extends to the depressions of Farafra and Bahariya. The surface of the plateau slopes northward and ends in a great depression, some parts of which are below sea level (e.g., Siwa and Qattara).

There are various places in the depressions of the Western Desert where long parallel lines of high sand dunes extend in a north-south direction, sometimes for great distances. The most famous sand dunes are the eastern series on the line of the oases, extending about $700\,\mathrm{km}$ to the south of Kharga.

The Bahariya oasis is a large natural excavation. It is entirely surrounded by escarpments and has a large number of isolated hills within the depression. The general shape of the depression is oval, with its major axis running northeast and with a narrow blunt extension at each end. It is situated on the stable-unstable shelf hinge; its principal structural feature is a very broad anticlinal fold, with its axis trending southeasterly from Gebel Ghorabi in the north and seeming to continue south to include the Farafra structure. The topography of the Bahariya oasis is characterized by a large number of hills within the depression. The most

Fig. 6.11 A geological map of Bahariya Oases, Western Desert, Egypt (Ball and Beadnell 1903)

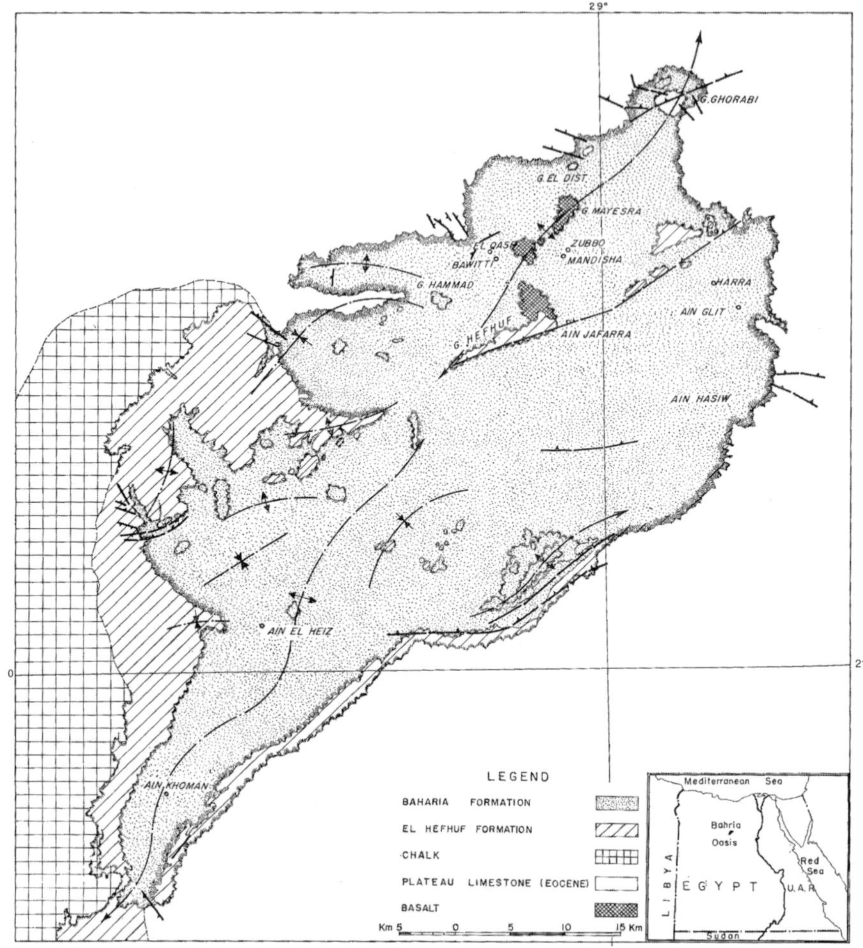

strongly marked group of hills extends in a straight northeast direction, cutting the oasis in two. The largest hill of the range is Gebel Hefhuf, a narrow, ridge-like hill of limestone (Fig. 6.11).

The basement in the Bahariya oases was reached at a depth of 1,825 m, where the Paleozoic sediments are succeeded unconformably by the Mesozoic and by later sediments of shallow marine deposits that are relatively thin.

6.4.1.1 Stratigraphy of the Sedimentary Section of the Bahariya Oases

The stratigraphic rock units that make up the geology of the Bahariya oasis of the Western Desert of Egypt are described in chronological sequence:

5. **Dolerite intrusions**—mainly of laccolithic dolerites, which cap Gebel Mendisha and Gebel Maysera and are intruded as sill bodies in the Cretaceous rocks of Gebel Hefhuf.

4. **The limestone plateau**—mainly of nummulitic limestones of the Lower Middle Eocene age, is unconformably overlying the Bahariya formation both in the wall of the escarpment and in the isolated hills within the depression.

3. **Chalk formation**—comprised of chalky limestones that belong to the chalk rock unit and are unconformably overlain by the Hafhuf Formation.

2. **El-Hafhuf formation**—mainly brown to gray crystalline limestones in its lower part and variegated shales and sandstones in its upper part. It occurs in the series of hills in the center of the depression, situated more or less along a fault line that crosses the oasis diagonally.

1. **Bahariya Formation**—mainly of sandstones and variegated shales, forming the floor of the oasis and parts of the wall of the escarpment. In the north-

ern part, the sandstones are overlain immediately by the Eocene plateau limestone. In the south they are followed by El-Hafhuf limestones and variegated shales. In the isolated hills within the depression of the Baharyia, sandstones and variegated shales are capped either by the Eocene plateau limestone or by basalts or dolerites. The composite thickness of the exposed part of this unit is about 300 m.

6.4.1.2 El-Bahariya Petroleum Exploration Well # 51

In the early 1950s, the American Sahara Petroleum Co. (SAPETCO) drilled an exploration well in the Bahariya oasis, but it was plugged in 1956 because of a political crisis within Egypt. The well was drilled mainly in sandstone with little silt and shale intercalations (Chart 6.2); it reached the basement rock (basalt) at 1,841.6 m. Figure 6.12 shows a cross section from Matruh on the northwestern coast to the southeast at Assiut on the Western Bank of the Nile River, passing by the Qattara depression and the Bahariya oases. The formation tops are defined by gamma and resistivity logs and intervals and can be summarized as follows:

Cenomenian : 00.00–705.6 m
Pre-Cenomenian: 705.6–1,365.5 m
Cambrian : 1,365.5–1,822.7 m
Pre-Cambrian : 1,822.7–1,841.5 m (TD)

6.4.2 Geological Results of the Petroleum Exploration of the Assiut-Kharga Well

The Assiut-Kharga deep exploratory well, drilled by the Egyptian General Desert Development Authority (EGDDA), lies nearly midway on the main asphalt road connecting Assiut city on the River Nile valley and the Kharga oasis (Fig. 6.13). It is the only deep well in this area and is about 120 km southwest of Assiut. Its latitude and longitude are 26°30′N and 30°54′E, respectively. The total depth reached at 1,140.5 m spudded at the lower Eocene and was abandoned in the lower Senonian due to pipe stuck (Chart 6.3).

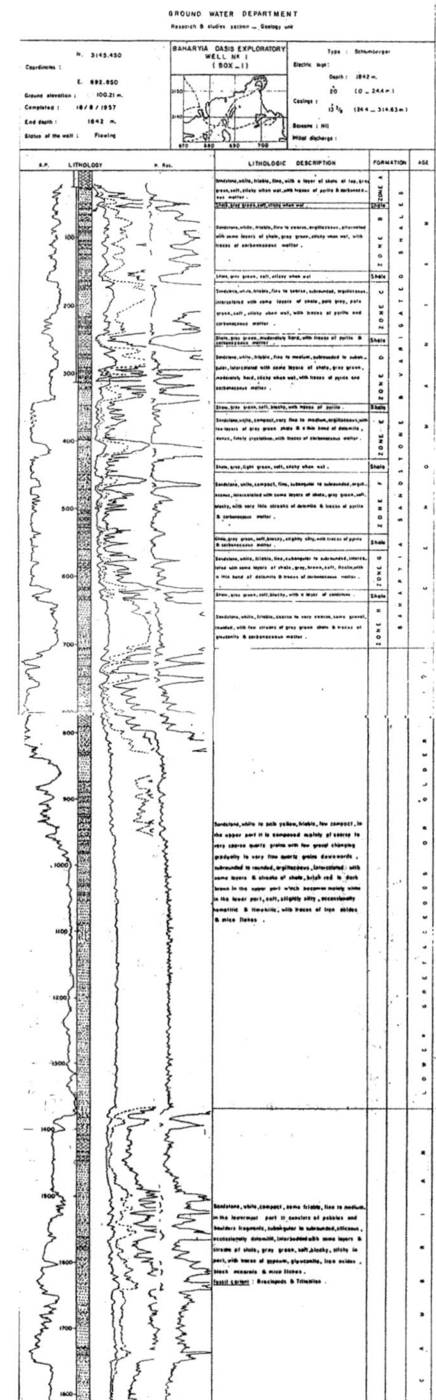

Chart 6.2 A composite log of Bahariya well # 1

6.4.2.1 Discussion

Horizontal Eocene beds outcrop at the well site area and are mainly nonclastics, hard limestone, partially

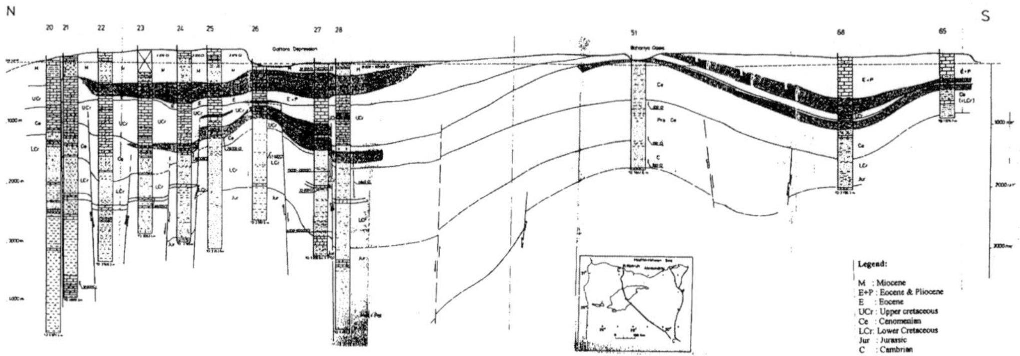

Fig. 6.12 Generalized Cross Section: Mersa Matruh – Qattara Depression – Bahariya – Assuit

dolomitized, gray to yellowish, and are intercalated by flint nodules. The beds are similar to those exposed in the Gebel Drunka section to the southwest of Assiut. It was estimated that the thickness of the whole Eocene section (particularly the Lower Eocene) in the vicinity of the well site ranges from 300 to 400 m.

Bishay and Tawfik (General Petroleum Co. 1965) described similar beds exposed to the east of Assiut on the other side of the Nile which are differentiated by two formations: a lower nonfossiliferous unit of 180 m of chalk marly limestone with intercalations of flint known as the "Assiuty chalk formation." The upper unit is nummulitic crystalline limestone 350 m thick, known as the "Manfalout formation." To the northwestern side of the Kharga depression, a section of Gebel El-Tir made up of upper Cretaceous beds is overlain by 10 m of lower Eocene, and is estimated to be 138 m in thickness.

No detailed geophysical investigations have been made to cover the area between Assiut and Kharga. A magnetic survey was carried out in the Wadi El-Assiuty area northeast of Assiut; it was concluded that the basement complex is a NE-SW trend of a local high or a dome structure complicated by faulting. The thickness of the sedimentary section overlying the basement complex ranges from 500 m in the highs to 3,000 m in the lows.

The stratigraphic description and facies analysis of the subsurface section are mainly based on ditch samples which were collected every one meter, and on five core samples recovered from the following depths: 117–883 m, 378–384 m, 551–557 m, 799–800 m, and 882–888 m, respectively.

The stratigraphic boundaries were determined at the General Petroleum Company (Table 6.6; Tawfik, personal communication).

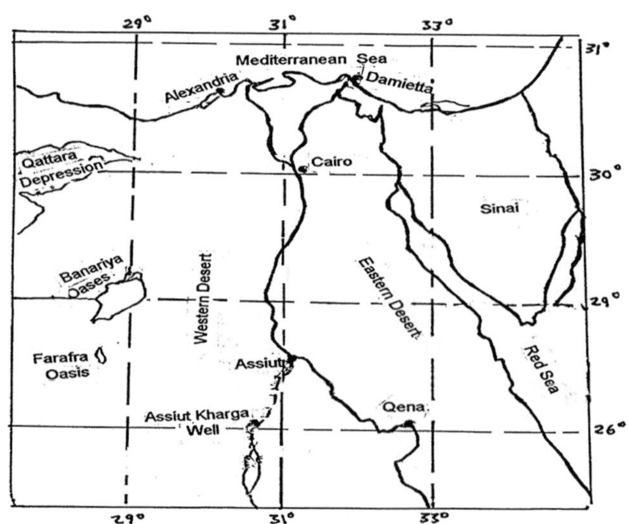

Fig. 6.13 Location of Assiut-Kharga Well, Egypt

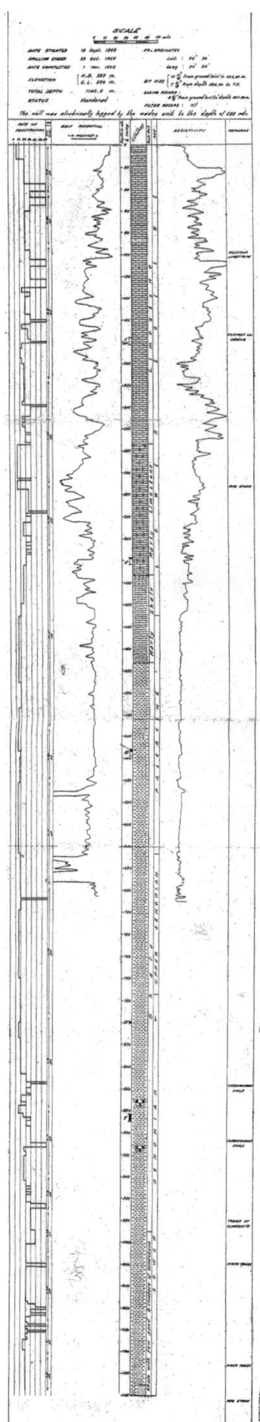

Chart 6.3 Assiut-Kharga composite well log

Table 6.6 Stratigraphical units in Assiut-Kharga well

Stratigraphic unit	Depth from Derrick		Top to sea	Thickness
	From (m)	To (m)	level (m)	(m)
Lower Eocene	000	430	+320	430
Paleocene	430	648	−110	218
Upper Senonian	648	775	−328	127
Lower Senonian	773	1,140.5	−455	365.5

deep borehole located in the northernmost part of the Kharga depression.

The uppermost beds comprise the Upper Senonian variegated shales about 40 m thick, overlying the Nubia sandstone 715 m thick, and are bottomed by weathered basement complex, drilled in granite wash for 20 m.

The Nubia sandstone at Gebel Um El-Ghanayem is 40 m thick and is overlain by the Paleocene shales and chalk that reach 106 m in thickness; the topmost Nummulitic limestone is 94 m thick at the Lower Eocene.

The Nakb Assiut escarpment (or Assiut pass) to the far east of the Kharga oases chronologically comprises Upper Senonian, Paleocene, and Lower Eocene beds of 107.5 m, 143 m, and 36 m, respectively (Fig. 6.14; Assaad and Barakat 1965).

From the stratigraphic point of view, the Assiut well occupies a site on the fringe of a sedimentary basin that developed under the influence of continuous sedimentation during Eocene–Upper Cretaceous times. Increase in thickness is expected further northeastwards. The possibility of encountering marine Jurassic, Triassic, and upper Paleozoic sediments like those recorded by the deep exploration wells in the Western Desert is still realistic. The basin is the extension of the Nile Valley basin from Beni-Suef in the north to "nearby" Assiut in the south. It seems to offer good oil prospects, because black shales are well developed in the Lower Eocene, Paleocene, and Upper Cretaceous sediments, and could act as the mother rock of petroleum hydrocarbons, as in the case of the Paleocene shales of Libya. Oil could be trapped in the structural highs of the fissured limestone of the Lower Eocene. Thus, further drilling is strongly recommended to delineate clearly the limits of the basin and define its facies changes both laterally and vertically.

6.4.2.2 Local Correlation

The stratigraphic formations encountered in Bahariya well #51 are correlated with those of the Um El-Kosour

6.4.3 The Nile Delta of Egypt*

The Nile Delta stretches 170 km in a north-south trend, attaining a base of 220 km along the Mediterranean

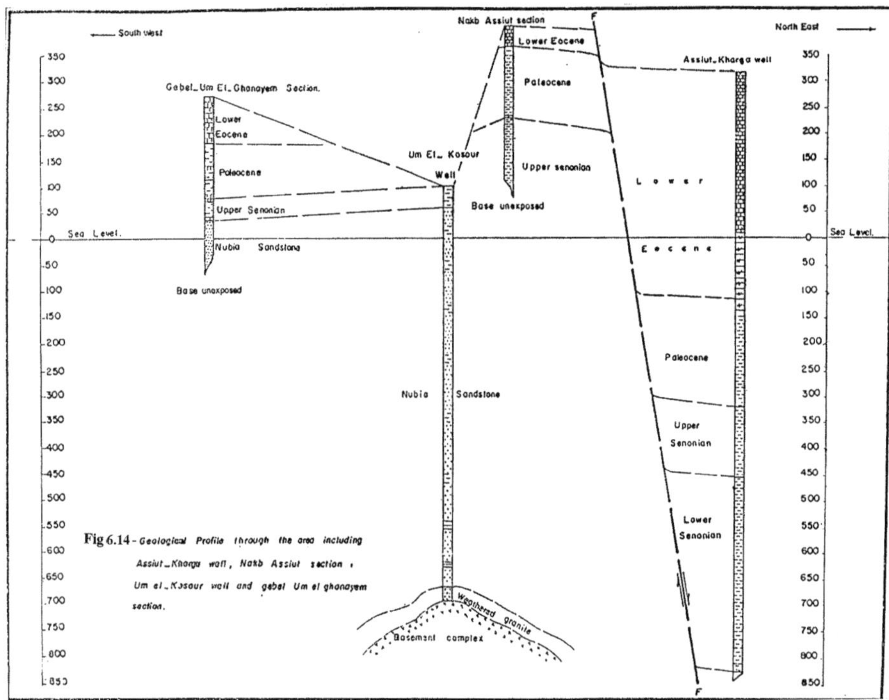

Fig. 6.14 Geological profile through the area including Assiut-Kharga well, Nakb Assiut section, Um el-Kosour well and gebel Um el-ghanayem section

Sea; it covers an area of $22,000\,km^2$ onshore, and about an equal area in the offshore continental[*] shelf and slope (Metwalli et al., 1978; Fig. 6.15). The Nile Delta represents a consequence of the conflict between the River Nile and the Mediterranean Sea. An understanding of the interaction of the Nile Delta together with the southern Mediterranean, provides the key that unlocks the possibilities and vision of significant petroleum reserves hiding in the delta area (onshore and offshore).

The Nile Delta, like the Niger Delta, is a product of approximately equal contributions from riverine and wave forces (fluvial-marine interaction). Continuous beach and beach-ridge formations fringe the coastlines of both deltas but are most extensively developed along the coast of the Nile Delta.

The episodical fracturing of the overlying salt deposits of the Miocene clastics permitted the hydrocarbons to migrate towards the continental shelf and the Nile Delta body mass. The high pressure source areas might represent abnormally pressured hydrocarbon compartments that are the source of lateral and vertical migration of hydrocarbon gasses into the normally pressured Pliocene clastics, which form significant reservoir-traps in the northern Delta offshore area.

The Tertiary Nile Delta is not a true basin but later overprints onto other basin types which represent the interaction of the coastal setting of the Mediterranean Sea. It is temping to visualize the Nile Delta as an echo of the rifting of the Red Sea and the Gulf of Suez. It might be a result of a triple tectonic junction (rifting) from different tectonic sources in time and space, reflecting: (1) the Red Sea, the Gulf of Suez, and Gulf of Aqaba rifting trends; (2) the Western Desert tectonics echo; and (3) the acquired post-rifting Atlantic margin from the north, the "Tythes and the Mediterranean ridge" (Metwalli 2000).

The Nile Delta area produces very high hydrocarbon gases (biogenic and thermogenic), yet crude oil prospects may be promising in the Pre-Miocene offshore concessions close to the delta's depocenter and consequently, in the Nile Delta continental cone and the Levant platform areas (Metwalli et al., 1978; Fig. 6.16).

Saiid (1981) reported that the River Nile owes its origin to a major tectonic event which affected the Mediterranean Sea in Late Miocene time. Crustal

---

[*] By M.H. Metwalli, G. Philip, El-Sayed, A., and A. Youssif, vol. 28, No. 3, Warsawa, 1978

Fig. 6.15 A location map of Northern Nile Delta, Egypt, (Metwalli et al., 1978)

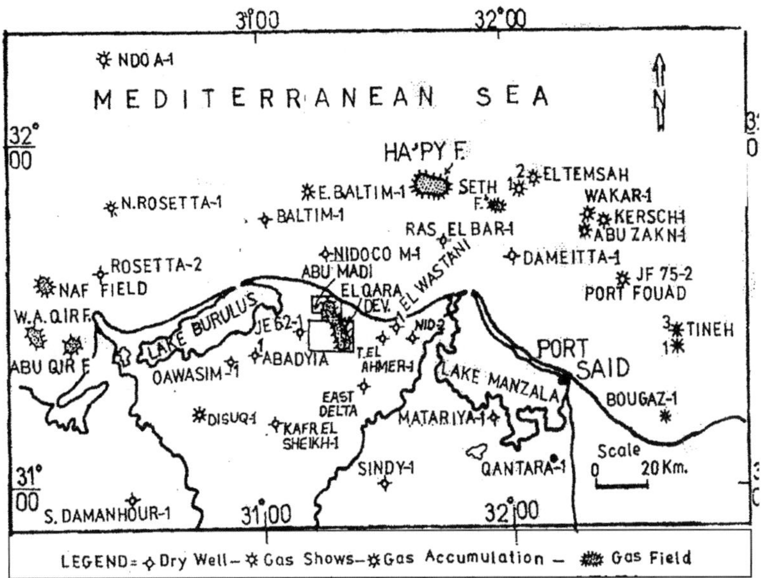

movements obstructed the connection between the Atlantic Ocean and the straits of Gibraltar. The interrupted inflow from the Atlantic Ocean led to the gradual desiccation of the Mediterranean Sea, with subsequent development of extensive thick salt and evaporitic deposits (the Rosetta formation).

6.4.4 The Gulf of Suez Region, Egypt

The Gulf of Suez lies within the stable belt of Egypt. It runs in a NW-SE direction, following the Erythrian trend of faulting, and forms an elongated depression separating the massifs of central Sinai from those of the backbone of the Eastern Desert (Fig. 6.17; Metwalli et al., 1978).

The coastal strip along the Gulf of Suez region that lies between the waters of the Gulf and the major African trending faults represents one of the most promising areas for the finding of future oilfields. Saiid (1962) stated that bordering the Gulf of Suez depression on both sides are two marginal faulting zones. Situated within the stable belt of Egypt, the Gulf of Suez has been an active zone of subsidence (or trough) throughout its geological history. Great thicknesses of sediments accumulated in the trough, and practically all the stages from the Paleozoic to the Recent are represented by regional faulting zones, usually marked by lines of high vertical escarpments on the upthrown sides. The successive Paleozoic-Mesozoic and Tertiary movements that affected the Gulf area have resulted in the most intensively faulted area in the world.

6.4.4.1 The Miocene Reservoir of El-Morgan Offshore Oil Field

The startigraphic section in the Gulf of Suez region ranges in age from Paleozoic to Recent, but most of the drilled wells ended in Tertiary rocks; whereas the Miocene evaporates and the underlying Miocene clastics are the main sequence penetrated. Several inshore and offshore oil fields were drilled, of which the Morgan offshore oil field is presented as a type of producing well.

The oil field of Morgan is an offshore oil field, SW of the Gulf of Suez petroleum province of Egypt; it embraces an area of $46\,\mathrm{km^2}$ and is situated in faulted blocks of reservoir rocks. It was a traditional concept that hydrocarbon accumulations in the Gulf of Suez region were confined within the Tertiary sediments, especially Miocene. However, the pre-Miocene deposits were not fully penetrated, although detailed studies on the geologic, tectonic setting and the mode of salt movement, together with detailed structural analysis of the nearby areas, might help in exploring unknown hydrocarbon reserves underlying the Miocene and in deeper reservoirs that were normally capped by thick Miocene salt deposits (Metwalli

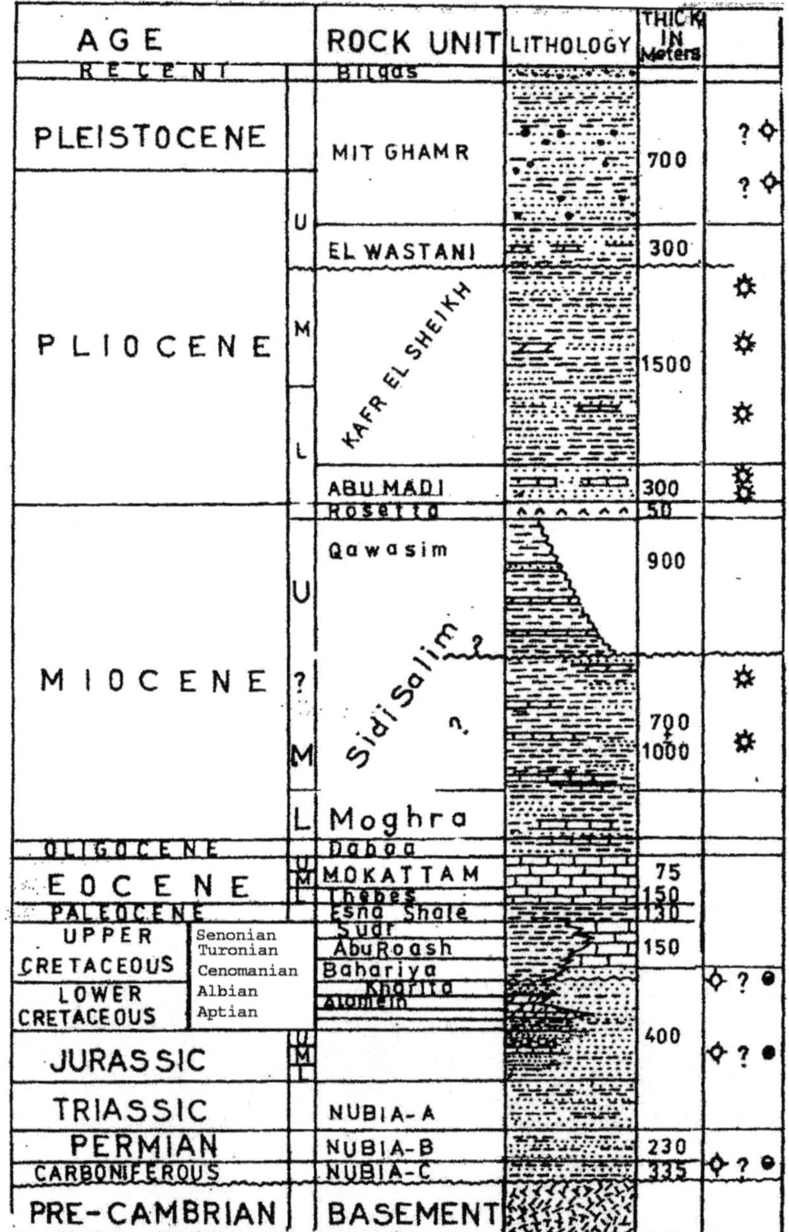

Fig. 6.16 Generalized hydrocarbon-bearing litostratigrapic column of the Nile Delta with inferred old tertiary and pre-tertiary sequences (Modified after Schlumberger, 1964)

et al. 1978). The Morgan field lies on the eastern bank rich of coral reefs that give its Arabic name, "El-Morgan".

Metwalli et al. (1976) referred to the tectonic setting of the field, with special reference to salt structures, and to its movements that affect to some extent the hydro-carbon accumulations as well as the source-reservoir relations. Figure 6.19 (Metwalli et al. 1978) shows the lithostratigraphic section of the Morgan oil field where the Miocene evaporite and shale deposits overlie differ-ent reservoir rocks, the most important of which are:

(a) **Zeit formation** (Late Miocene-Pliocene)—Lies on top of Miocene deposits, mainly of evaporites

Fig. 6.17 A Location Map of Gulf of Suez showing the Morgan Oil Field and the position of cross line C-C′, (Metwalli et al. 1978)

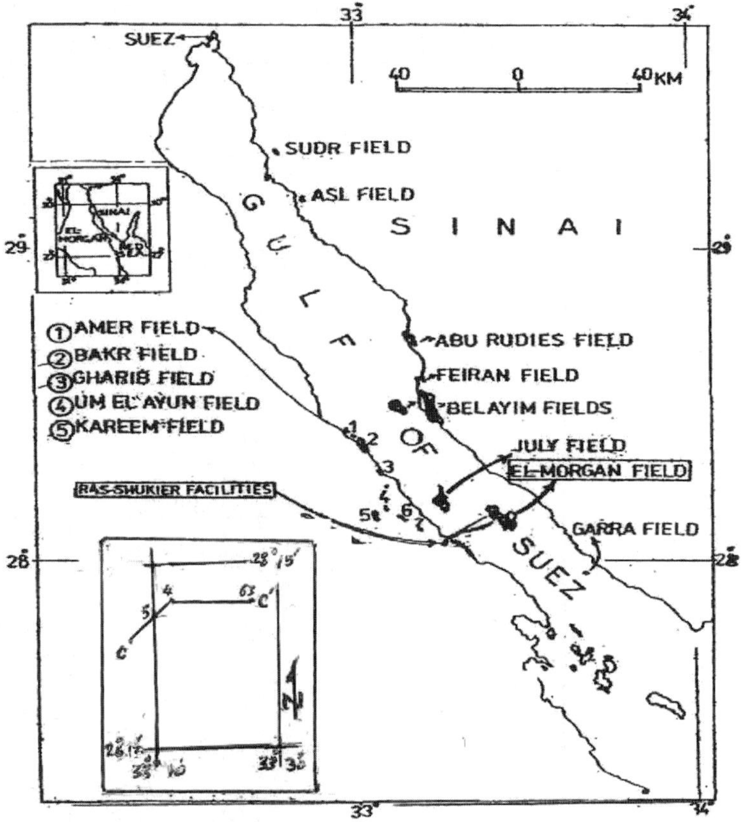

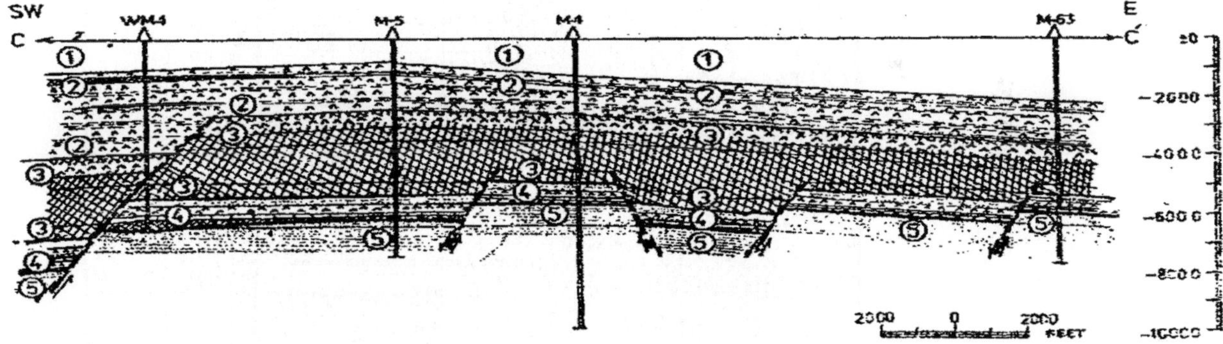

Fig. 6.18 A structural cross section across C-C′ (see Fig. 6.17/location of sec. CC′) (Metwalli et al. 1978)

and shale intercalations, directly underlying the post-Miocene clastic deposits. Hydrocarbons in the Zeit formation at the El-Morgan oil field are not recorded; however, it is oil bearing at the Belayim onshore oil field.

(b) **South Gharib formation**—Is oil producing in many fields of the Gulf of Suez petroleum province, characterized by a remarkable increase in the thickness of evaporites (mainly rock salt) of proper lagoon facies, where salt bulges are defined in some locations and cause many problems in interpreting seismic data. No traces of hydrocarbons are known in the South Gharib formation at the El-Morgan oil field; however, it is oil producing at the Belayim onshore and Bakr oil fields. The formation acts as a sealing unit

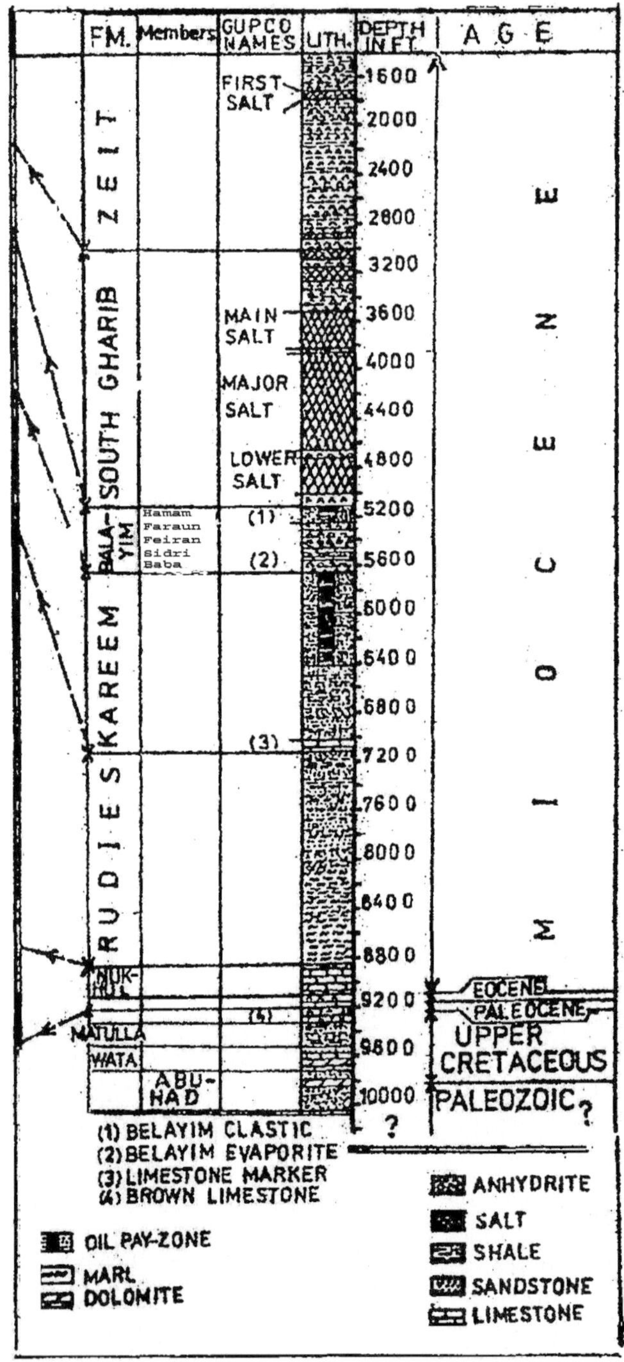

Fig. 6.19 A simplified Lithostratigraphic Log in Morgan Oil Field, (Metwalli et al. 1978)

that prevents the oil migration of the underlying Belayim and Karim formations updip at the El-Morgan oil field.

(c) **Belayim formation**—Is about 1,000 ft (~333 m) in thickness, consists of evaporates and interevaporite marls. It is oil producing in many fields in the Gulf of Suez petroleum province, e.g., the El-Morgan, Bakr, and Gharib offshore, as well as the Belayim onshore oil fields. The topmost member is known as the Belayim clastics, which are collectively known as the Belayim evaporites.

(d) **Kareem formation**—Attains a thickness of about 1,400 ft (~400 m) at the El-Morgan field and is composed of calcareous arkosic sandstone with thin shale interbeds and minor evaporite intercalations. The sand and sandstone of the Kareem formation form the pay zone, producing oil of about 30° API; it is also productive in both the Kareem and the Belayim onshore oil fields. At the Gharib field, it reaches about 800 ft ($\sim$250 m) in thickness of clastics, and is interbedded by anhydrite deposits and occasional limestone.

(e) **Rudies formation**—Attains a thickness of about 2,550 ft ($\sim$400 m) at the Suez Gulf region and consists mainly of sandy clays. The clays are highly calcareous with abundant planktonic foraminifera (Globigerina marl). The Rudies formation is oil producing at the Balayim onshore and offshore oil fields, and the Feiran and Kareem oil fields.

6.5 Economic Aspects of Salt Structures

The tectonic outline in areas of thick salt deposits should be carefully studied to define the different types of salt dome structures buried deep underneath the ground, for the eventual storage or disposal of high-level solidified radioactive waste. These hazardous waste materials are dry, mainly impervious to water, and not associated directly with useable sources of groundwater wastes. Large caverns can be mined out at depths of up to 300 m where two-thirds of the salt can be dissolved by hot water and removed from the salt dome structure with only slight deformation of the support pillars (by pumping hot water). Then the cavern can be lined with impervious material, e.g., asphalt.

Salt can be extracted either by conventional underground mining or by solution mining, in which the salt is dissolved in the dome by controlled circulation of fresh water and subsequent evaporation of saturated brines that are pumped or flowed to the surface.

Taking into account the accumulation of hydrocarbons, the economic importance of salt structures can be described in terms of their occurrence and distribution, accessibility, and productivity, as well as other factors. In general, the mining of salt and sulfur and the quarrying of limestone can be carried out from the cap rock and/or from salt pillows and salt domes.

References

Aliev, M., Ait Laussine, Said, A. et al. (Sonatrach, 1971) Geological Structures and Estimation of Oil and Gas in the Sahara of Algeria: Spain, Altamira-Rotopress, S.A., 255 pp.

Assaad, F. (1972) Contribution to the study of the Triassic formations of the Sersou-Megress Region (High Plateaux) and the area of Daia M'Zab (Saharan Platform), No. 84(B3), Eighth Arab Petroleum Congress, Algeria, 1972.

Assaad, F., (1981) A further geologic study on the Triassic formations of North-Central Algeria with special emphasis on Halokinesis, Journal of Petroleum Geology, Vol. 4, pp. 163–176.

Assaad, F.A. (1988) Hydrogeological aspects and environmental concerns of the New Valley, Region, western desert, Egypt, with special emphasis on the Southern Area, Environmental Geology Vol. 12, No. 3, 141–161.

Assaad, F.A, and Barakat, M.G. (1965) Geological results of the Assiut Kharga well, Egypt, the Journal of Geology, UAR, Vol. 9, No. 2, pp. 81–87.

Bishay, F. and Tawfik, A.E. (1965) in Journal of Geology, UAR, Vol. 9, No. 2, pp. 81–87.

Burollet, P.F. et al. (1978) The geology of the Pelagian Block: the margins and basins off southern Tunisian and Trilopitania, in Nari, A.E.M., Kanes, W.H., and Stehli, F.G., eds., The Ocean Basins and Margins: New York, Plenum Press., v. 4B, pp. 331–359.

Klett, T.R., Ahlbrandt, T.S. Schmoker, J.W., and Dolton, G.L., (1997) Ranking of the world's oil and gas provinces by known petroleum volumes: U.S. Geological Survey Open-File Report 97-463.

Kun, N.D. (1965) The Mineral Resources of Africa, Columbia University in the City of New York, N.Y., Elsevier Publishing CO., 52 Vanderbilt Av., New York, N.Y. 10017, (Figs. 12, 58,123 & 125).

Magoon, L.B., and Dow, W.G. (1994) The petroleum system, in Magoon, L.B. and Dow, W.G, Eds, The Petroleum System – From Source to Trap: American Association of petroleum Geologists, Memoir 60. pp. 3–24.

Metwalli, M.H. (2000) A New Concept for the Petroleum Geology of the Nile Delta, A.R. Egypt, Geol. Dept. Fac. Science, Cairo, Egypt, 5th., International Conference on the Geol. of the Arab World, Cairo, Univ., pp. 713–734.

Metwalli, M.H., Philip, G., Wali, A.M.A. (1979) Repeated Folding and its-significance in northern Western Desert , Petroleum Province , Egypt, Vol. 29, No. 1, pp. 133–150; Acta Geologica, Polonica, Poland, Warsawa.

Metwalli, M.H., Philip, G., Youssef, A.A.E. (1978) El Morgan Oil Field as a Major-Blocks Reservoir masked by thick Miocene Salt; Vol. 28, No. 3; Acta Geologica Polonica, Poland, Warsawa.

Metwalli, M.H., Philip, G., Youssef, A.A.E. (1982) Petrographic characteristics of the oil-bearing formations in El-Morgan oil field "Gulf of Suez Petroleum Province", A.R. Egypt. Acta Geologica Academiae Scientiiarum Hungaricae, Vol. 25 (3–4), pp. 275–295.

Metwalli, M.H., Philip, G., Youssef, A.A.E. (1981) El-Morgan oil field crude oil and cycles of oil generation and accumulation in the gulf of Suez petroleum province, A.E.R., Egypt;

Acta Geologica Academiae Scientiarum Hungaricae, Vol. 24 (2–4), pp. 369–387.

Michel R (1969) Petroleum developments in North Africa. Am Assoc Petrol Geol, in AAPG 1970; Bull 54(8):Fig. 1.1

Montgomery, S., (1993) Ghadames Basin of north central Africa; Stratigraphy, geologic history, and drilling summary, Petroleum Frontiers, Vol. 10, No. 3, 51 p.

Persits, F. and others (1997) Maps showing geology, oil and gas fields and geologic provinces of Africa: U.S. Geological Survey Open-File Reopoport 97-470A, CD-ROM.

PetroCinsultants (1996b) PetroWorld 21: Houston, Tex., PetroConsultants, Inc. [database available from Petroconsultants, Inc., P.O. Box 740619, Houston, TX 77274-0619].

SN RÉPAL (1961) Les séries Permo-Triassiques dans le Nord Sahara. Études Pétrographique du cycle Détritique. Etudes du Cycle Salifére (texte et planches).

Said, R. (1962) The Geology of Egypt, Cairo University, Egypt, UAR, Fig. 13, Elsevier Amsterdam, Publishing CO.

Stoica, I., and Assaad, F.A. (1981) Geological Studies on the Triassic Reservoir of Condensate and Gas at the Hasssi R'Mel Field (Internal Report, Sonatrach).

Werwer, A.M., (1973) Hydrogeological Study of the Wadis El-Ajal, Shatti, Traghen, Libya, by REGWA, Cairo, Egypt.

Websites:

– Map of Algeria/AAPG: [mem-algeria.org/hydrocarbons/geology.htm]
– Klett, T.R. (9/23/2006): [http://greenwood.cr.usgs.gov/pub/bulletins/b2202.bl/

(1) Total Petroleum Systems of the Illizi Province, Algeria and Libya: [Pubs.usgs.gov/bul/b2202-a/b2202-bso.pdf]

(2) Total Petroleum Systems of the Grand Erg/Ahnet Province, Algeria and Morocco: [Pubs.usgs.gov/bul/b2202-b/b2202-bso.pdf]

(3) Total Petroleum Systems of the Trias/Ghadames Province, Algeria, Tunisia, and Libya: [Pubs.usgs.gov/bul/b2202-c/b2202-bso.pdf]

Appendix

Geological Time Scale

PHANEROZOIC

CENOZOIC

AGE (Ma)	Period	Epoch		Stage	AGE (Ma)
0	Quaternary	Holocene			
		Pleistocene			
	Neogene	Pliocene	L	Gelasian	1.81
				Piacenzian	2.59
			E	Zanclean	3.60
					5.33
		Miocene	L	Messinian	7.25
				Tortonian	
					11.61
			M	Serravallian	13.65
				Langhian	15.97
			E	Burdigalian	20.43
				Aquitanian	23.03
	Paleogene	Oligocene	L	Chattian	28.4
			E	Rupelian	33.9
		Eocene	L	Priabonian	37.2
				Bartonian	40.4
			M	Lutetian	48.6
			E	Ypresian	55.8
		Paleocene	L	Thanetian	58.7
			M	Selandian	61.7
			E	Danian	65.5

MESOZOIC

AGE (Ma)	Period	Epoch	Stage	AGE (Ma)
				65.5
70	Cretaceous	Late	Maastrichtian	70.6
			Campanian	
			Santonian	85.5 / 85.8
			Coniacian	89.3
			Turonian	93.5
			Cenomanian	99.6
		Early	Albian	112.0
			Aptian	125.0
			Barremian	130.0
			Hauterivian	136.4
			Valanginian	140.2
			Berriasian	145.5
	Jurassic	Late	Tithonian	150.8
			Kimmeridgian	155.7
			Oxfordian	161.2
		Middle	Callovian	164.7
			Bathonian	167.7
			Bajocian	171.6
			Aalenian	175.6
		Early	Toarcian	183.0
			Pliensbachian	189.6
			Sinemurian	196.5
			Hettangian	199.6
	Triassic	Late	Rhaetian	203.6
			Norian	216.5
			Carnian	228.0
		Middle	Ladinian	237.0
			Anisian	245.0
		Early	Olenekian	249.7
			Induan	251.0

PALEOZOIC

AGE (Ma)	Period	Epoch		Stage	AGE (Ma)
	Permian	Lopingian		Changhsingian	251.0 / 253.8
				Wuchiapingian	260.4
		Guadalupian		Capitanian	265.8
				Wordian	268.0
				Roadian	270.6
		Cisuralian		Kungurian	275.6
				Artinskian	284.4
				Sakmarian	294.6
				Asselian	299.0
	Carboniferous	Pennsylvanian	Late	Gzhelian	303.9
				Kasimovian	306.5
			Middle	Moscovian	311.7
			Early	Bashkirian	318.1
		Mississippian	Late	Serpukhovian	326.4
			Middle	Visean	345.3
			Early	Tournaisian	359.2
	Devonian	Late		Famennian	374.5
				Frasnian	385.3
		Middle		Givetian	391.8
				Eifelian	397.5
		Early		Emsian	407.0
				Pragien	411.2
				Lochkovian	416.0
	Silurian	Pridoli			418.7
		Ludlow		Ludfordian	421.3
				Gorstian	422.9
		Wenlock		Homerian	426.2
				Sheinwoodian	428.2
		Llandovery		Telychian	436.0
				Aeronian	439.0
				Rhuddanian	443.7
	Ordovician	Late		Hirnantian	445.6
					455.8
		Middle		Darriwilian	460.9
					468.1
		Early			471.8
				Tremadocian	478.6
	Cambrian	Furongian			488.3
				Paibian	501
		Middle			513
		Early			542.0

PRECAMBRIAN

AGE (Ma)	Eon	Era	Period	AGE (Ma)
				542
600	Proterozoic	Neoproterozoic	Ediacaran	630
			Cryogenian	850
			Tonian	1000
		Mesoproterozoic	Stenian	1200
			Ectasian	1400
			Calymmian	1600
		Paleoproterozoic	Statherian	1800
			Orosirian	2050
			Rhyacian	2300
			Siderian	2500
	Archean	Neoarchean		2800
		Mesoarchean		3200
		Paleoarchean		3600
		Eoarchean		Lower limit is not defined

Chapter 7

The Petrodata DataBase System

7.1 Scope

Part II of this volume is offered as a model representation of lithostratigraphic correlation charts of sedimentary formation in different territories; the geology of the Algerian Saharan Platform covers different structural basins and was primarily selected as an approach to establish a computerized regional lithostratigraphic correlation chart of the Arabian Maghreb in northwest Africa.

In 1970, the Algerian National Petroleum Organization (Sonatrach), assigned a group of Algerian geologists and engineers to the International Petrodata Incorporation (IPI) of Canada to help build the Petrodata DataBase Retrieval System, the most sophisticated computerized system in Africa and the Middle East. The Petrodata System was exclusively established by the Exploration Directorate of Sonatrach (Societé Nationale De Transport et De Hydrocarbures) to build up a database for electronically storing geological, geophysical, and production petroleum data to be processed later for further applications.

In the late 1970s, the author, being in charge of the Petrodata Project, established a computerized lithostratigraphic correlation chart that covered different regions of the Algerian Saharan Platform. A lithostratigraphic chart should be continuously updated and modified by different academic and government institutions. The African Petroleum Committee, which had been established several decades ago, had apparently stopped its activities, probably for political reasons. However, the cultural and media attaché of the Algerian Embassy in Cairo demonstrated the Algerian government's special interest in the subject.

7.2 The Petrodata System

7.2.1 Historical Aspect

The International Petrodata Incorporation (IPI) of Canada used a Sigma 7-frame computer with permanent disc storage for editing, updating, extracting, and selecting the geological, geophysical, and production data of petroleum fields; the data were computerized in digital form for statistical analysis and for computer applications.

The author was assigned to provide liaison with the Exploration Districts and research groups; to supervise and manage the quality of the input general well data; the geological, geophysical, drilling, and production data, etc.; as well as to define and formulate retrieval requests for computer extraction procedures; supervise corrections and improvements of input data; and to refine the edit and cross check functions of the maintenance subsystem.

The IPI firm had provided a complete set of volumes for explanatory documentation purposes and for training different qualified Algerian engineers and geologists to carry out satisfactorily the Petrodata project through a certain period of time; however, the initial objectives of the project, applying all the components of the system, were not completed because much time was lost at the beginning of the project transferring the

F.A. Assaad, *Field Methods for Petroleum Geologists*,
© Springer-Verlag Berlin Heidelberg 2009

utility programs and subroutines to the Sigma 7 computer system and then later, in 1982, updating to an IBM/VM/370 mainframe computer system.

7.2.2 Classification of the Petrodata System

Certain procedures are required to operate the Petrodata system for a particular territory. They comprise three sequential subsystems: maintenance, retrieval, and application. Fig. 7.1 is a flow diagram of the Petrodata system.

7.2.2.1 The Maintenance Subsystem

The maintenance subsystem is an essential primary step to input, validate, and update data until a severity code is accepted and stored in the database (DB); all data in the database are indexed by the database index feature that is available on the maintenance subsystem whenever the database is accessed.

The maintenance subsystem is the capstone source of the petrodata system, because it is responsible for storing geographical, geological, geophysical, and production well data for each specific petroleum field. It establishes a complete record of these data, which should be processed through a validation program that

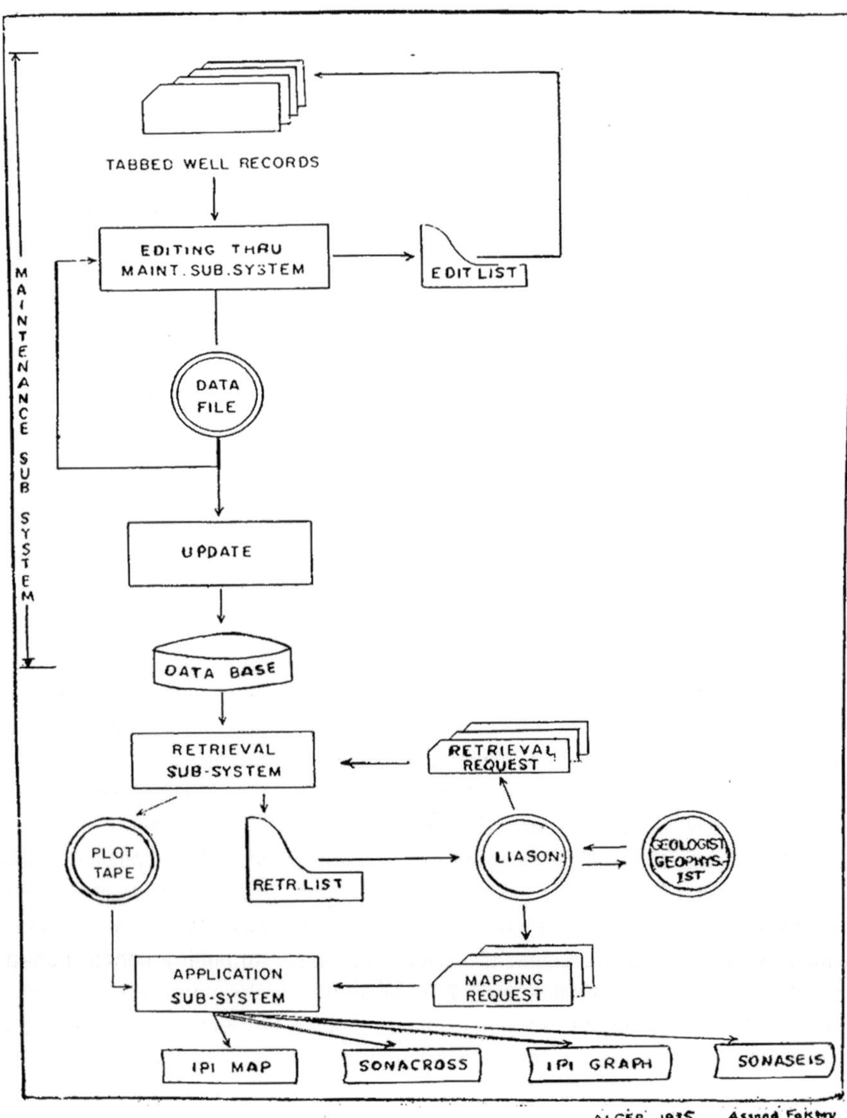

Fig. 7.1 Flow diagram – petrodata system

edits input data, and an update program that stores the final processed data as "clean data" in the database. The main functions required to build the database are:

(1) sorting data into a meaningful order;
(2) correcting incorrect data;
(3) validating all submitted data;
(4) defining data storage allocation; and
(5) storing data in a format suited for the retrieval sub-system.

Software program files were established in sequential steps to carry on the maintenance process.

The post-input sort file (**XSORT**) places the valid format type of the input data into the desired sequence; the extraction file (**XEXTRACT**) stores the validated input data into an update tape for further correction and crosschecks the validation process; the post-extraction sort file places all extracted data into proper sequence; the first stage validation (**XFSV**) examines each field individually to determine its proper data type within the logic limits; the second stage validation (**XSSV**) examines related fields to predefine card type formats.

The data subjected to validation is then merged with data on the update tape from the previous run; the output of the **XSSV** step may be cycled through the extraction-correction runs to reach the first stage update (**XUPDATE 1**) step that can be primarily stored on the DB, "depending on the user specifications for the data and the system-generated validation severity errors."

The second stage update file (**XUPDATE 2**) is responsible for placing the clean data selected for the DB onto predetermined storage volumes of the retrieval subsystem to access and create backup copies of the DB required in case of a system failure.

The main program files of the maintenance subsystem can be classified into systematic jobs: syntax, validate, generate, and run jobs. A total of 105 explanation and production wells were selected and validated and reside on the database in specified files as a test for further retrieval and application procedures. Several coding procedures were established to carry out the maintenance tasks.

(A) Assignment of Catalogue Numbers for Well Identification in Algiers

A catalogue number for well identification is an essential tool of the Petrodata system to define the location of a well and for further application of a computerized lithostratigraphic columnar section of the encountered formations located in a particular structural basin of the territory.

An example of a catalogue number with eight alphanumeric fields is given for an exploration well in a specific area in north Algiers and can be explained as follows:

E	35–1349	73	A-171	K9	Sonatrach	Ras El-Guentis	Rgt-1
(1)	(2)	(3)	(4)	(5)	(6)	(7)	(8)

Field (1) comprises only one alphabetic code (or letter) that defines the nature of the well: E, exploration; S, stratigraphic; C, core; x, extension; W, water; I, injection; U, underground stockage.

Field (2) defines the well location in a specific basin of the encountered territory and is given in six numeric digits: the first indicates either the north (3) or the south (4) of the territory. The second denotes the location of a zone or a structural basin. In the Algerian Saharan Platform, there are five well known basins which are given the following numeric digits: Tindouf basin (1); Reggane basin (2); Illizi basin-Tinreht-SE Grand Erg Oriental (3); Ahnet, Mouydir-Tademait (4); and Bechar-Timmimoun (5).
(**NB**: no specific basins are known in north Algeria).
The remaining four digits of **Field (2)** are a sequential number indicating wells drilled in a certain zone or basin, normally starting from the number 1000.

Field (3) contains two digits that define the year of preparation of the montage well report.

Field (4) consists of four alphanumerics for the permit or concession number of the encountered well.

Field (5) contains two alphanumerics denoting the geographic unit of the well.

Field (6) holds an alphabetic that provides the name of the company that owns the concession or the permit related to the drilled well.

Field (7) holds an alphabetic for the encountered field name.

Field (8) consists of five alphanumerics for the well's abbreviation.

(B) Formation Coding System

The stratigraphic level on top of the rock units is defined by geologists and geophysicists in order for the supervisor to locate the related formation coding system (see the stratigraphic chart on CD-ROM).

Table 7.1 shows the formation coding system, composed of three coding fields: the **first field** holds the three-digit formation code; the **second field** holds the four-character alphabetic formation code; and the **third field** holds a six-digit numeric code which is machine generated.

An example is given below for two well known marker beds ("D2" and Hz "B") in the Triassic deposits of the Algerian Saharan platform.

The stratigraphic level of a particular bed ("D2": a thin limestone bed on top of the Triassic reservoir and the bottom of salt bed S4—"Rhetian Stage") and another limestone bed (Horizon "B" or Hz "B") on top of the anhydrite and salt beds of the cap rock (S1 + S2 + S3) of the Triassic reservoir ("Hettangian stage"); both can be defined by a six-digit code:

(a) The first digit specifies the Epoch or Era and is given "1" for Paleozoic, "2" for Mesozoic, and "3" for Cenozoic.

(b) The second digit specifies the subsystem, e.g., Triassic: "1," Jurassic: "2," Cretaceous: "3"—for the Mesozoic.

(c) The third digit denotes the chronological sequence of the above subsystem, i.e., the sequence of a formation in the stratigraphic column, e.g., Lower Triassic: "1," Middle Triassic: "2," and Upper Triassic: "3;" the same is applied for Lower Jurassic (Lias), Middle Jurassic (Dogger), and upper Jurassic (Malm).

(d) The fourth digit specifies the stratigraphic level (D2) on top of the Rhetian stage (the uppermost stage of the upper Trias (or Keuper: "3"); the Hettangian stage (of code "1") is at the base of the Lower Jurassic (Lias: "1").

(e) The last two digits (the 5th and 6th) give the stratigraphic level of the related lithological series of the rock unit; the stage (Étage) is divided into 10 subdivisions ranging from 00 to 99. Two examples are given: (1) the top of the marker bed HZ-B (old classification: the top of the Triassic formation) attained the stratigraphic level of the related stage (Hettangian): "99."

Table 7.2 shows two marker beds in the Triassic reservoir formation and can be explained as follows.

Table 7.2 The six-digit numeric code of a stratigraphic level of a bed

Base horizon "B"	Top bed "D2"
Mesozoic Code(2)	Mesozoic Code (2)
Jurassic **Code (2)**	Triassic **Code (1)**
Lower Jurassic Code (1)	Up.Triassic Code (3)
Hettangian Code (1)	Rhetian Code (3)
HZ "B" Code (99)	Bed "D2" Code (99)
221199	213399

Horizon B "Hz B"—code (2) for Mesozoic, code (2) for Jurassic, code (1) for Lower Jurassic, code (1) for Hettangian stage, and code (99) for "Hz-B," being on the top of the stage; hence the formation code should be 221199.

Marker bed "D2", "update correlation"—correlated as on top of the Triassic reservoir formation: code (2) for Mesozoic, code "1" for Triassic, code "3" for Upper Triassic, code "3" for the Rhetian stage, and the last two digits "99," because "D2" is located on top of the stage; hence the six digit code for the marker bed (D2) should be 213399 (see Fig. 6.4).

(C) Assignment of Well Data Files

All information pertaining to a well is organized into five subfiles: (1) general well data, (2) lithology data, (3) land data, (4) production data, and (5) geophysical data. Each of these subfiles comprises a group of related card type formats (total of 99) that should

Table 7.1 Formation coding system

Formation top	Three digit formation code	Alphabetic formation code	Six digit formation code
KEUPER	182	KUPP	213399
RHETIAN	335	RHTN	213399
TRIAS	404	TRAS	213399
BASE HORZ "D2"	178	HZD2	213399
SALT (S4)	544	SAS4	213399
SHALEY SALT	545	ARSA	213399

be always referred to the eight digits of the catalogue number of the well. A manual is designed to help geologists tabulate the data in the proper formats that cover all the above subfiles.

Table 7.3 shows a tabulating file that includes a list of the most required card types. The following is an example of some of the input data formats.

a. **General well data** comprises several card types (Table 7.4a); the most important are the following:

– **Card type 01**—(Well location): geographic well location, Lambert and UTM coordinates (if any), surface and reference elevation, well depth, and initial final status.

Table 7.3 Tabulating cards file

CATALOGUE N°______________________________   TAB. DATE ______

WELL NAME & CODE ____

* Required card
05 Required also for this card

CARD N°	TAB'D By & N°	TYPE OF INFORMATION	CARD N°	TAB'D By & N°	TYPE OF INFORMATION
01	*	LOCATION	39	(normally used)	CORRECTION
02	*	IDENTIFICATION	40		LAND
03	*	PROD. ZONE	41		—"—
04	*	LOGS	42		—"—
05	(normally used)	PERFORATIONS	43		—"—
06	(normally used)	CORE SUMMARY	44		—"—
07	(normally used)	NET PAY	45		SP. CORE ANAL.
08	*	FORMATION TOPS	46		—"—
09	(normally used)	TROUBLE ZONE	49	(normally used)	PROD. TEST
10	*	POROUS ZONE	50	(normally used)	—"—
11	(normally used)	WATER ANALYSIS	51		FORM. VELOCITY
12	(normally used)	—"— —"—	52		P/T GRADIENT
13	(normally used)	OIL ANALYSIS	53		—"—
14	(normally used)	TUBING	54		—"—
15	*	REMARKS	56	(normally used)	GAS ANALYSIS
16	(normally used)	CONTACTS	57	(normally used)	—"—
19	(normally used)	D.S.T.	58		FIELD DATA
20	(normally used)	—"—	59		WELL CHARACTER.
21	(normally used)	—"—	60		SAMPLING COND.
22	(normally used)	SCHLUMBERGER TST.	61		VOLUMETRIC
23	(normally used)	—"—	62		—"—
24	05	CASING	63		RES. FLUID
25		BUILD–UP PRESS.	64		O/G DENSITY
26		—"—	65		OIL —"—
27		—"—	66		RES. SEP. TESTS
28	*	DRILLING MUD	67		HYDROCARBON ANAL.
29	(normally used)	CORE CONTROL	68		—"—
30	(normally used)	CORE ANALYSIS	69		GAS ANALYSIS
32		PROD. INDEX	70		BUBBLE POINT
33		—"—	71		FLUID SUMMARY
35	(normally used)	LITHOLOGY	72		SEP. TEST
36	(normally used)	—"—	73		STAGE SEP.
37		—"—			
38	(normally used)	—"—			

REPORTS USED
____
____
____
____
____
____

REPORTS MISSING
____
____
____
____
____

DATE COMPLETED ______

CHECKED BY ______

DATE TO KEYPUNCH ______

(shaded) NORMALLY USED CARDS

(blank) UNUSED CARDS

Table 7.4a General Well Data (Card Types 1, 2 & 3)

CARDS 1, 2, 3

LOCATION 1–1

CATALOGUE	NUMBER	EMLA A EMLE					B	COUY C COUX			D	E ELSO	F ELRE		G		H		I			J	CARD TYPE 01
		LOCATION						COORDINATES				ELEVATIONS			WELL DEPTHS			PFPO	INITIAL STATUS			FINAL STATUS	
		LAT			LONG		E O R W				SURFACE	REF	TOTAL DEPTH		PLUG BACK TOTAL DEPTH	C L A S S I F I C	FIELD CODE BICC				
		DEG.	MIN.	SEC.	DEG.	MIN.	SEC.		Y	X		U T M	L A M N	L A M S			DRILLER PERD	LOGGER PRDIA							

IDENTIFICATION 1–1

NUCA	ABRE	IDPU	DEBU	SOND	COMP	CARTE TYPE 02
NUMERO DU	CATALOGUE	ABREVIATION	IDENTIFICATION DU PUITS	DATES		
				DEBUTE	SONDE RETIREE	COMPLETE
				JOUR MOIS ANNEE	JOUR MOIS ANNEE	JOUR MOIS ANNEE

POTENTIEL INITIAL

NUCA	FEAD	FOPR	DUPO	DUSQ DUMM SDTE	PIHU	PIGA	GORZP	LEAU	SEEA	X CONVERSION	CARTE TYPE 03
NUMERO DU CATALOGUE	OBTURE ET ABANDONNE		ZONE PRODUCTRICE								
	JOUR MOIS ANNEE	CODE DE FORMATION	DIAMETRE DE DUSE X 1/64 DOUCE X MM	SOURCE	PI HUILE	PI GAZ	G.O.R	EAU	% SED. ET EAU		

– **Card type 02**—(Well identification): well abbreviation, type of well, year of the montage report, geographic grit, well drilling and completion dates.
– **Card type 03**—(Potential initial): production well data such as hydrocarbon production data and other related measurements.
– **Card type 04**—(Log data): logging specifications.
– **Card type 05**—(Perforation data): perforation intervals and related porous zones within the production formation, encountered tops, and volume of acid used.
– **Card type 06**—(Core summary): intervals and length of core samples recovered.
– **Card type 08**—(Formation tops): actual or projected and unconformity tops.
– **Card type 15**—Technical or geological remarks on other card types.

b. Lithology Well Data comprises three main card types (Table 7.4b):

– **Card type 35**—(Lithology description): for rock units and related drilling intervals.
– **Card type 36**—(Miscellaneous description): for rock features.
– **Card type 39**—(Correction data): submit correction through the validation process of the input data.

c. Land Data mainly cover royalties and areas of concessions or permits, and geophysical data such as seismic shot lines and seismic sections.

NB: The **Database** can not be created until at least the second validation update stage is completed and the severity error does not exceed the allowed level (maximum severity of two). Better quality results are obtained, because of cross checking operations of the maintenance subsystem.

7.2.2.2 The Retrieval Subsystem

There are special factors that should be taken into account when determining the storage allocation, such as potential growth and type of data retrieved. The data are stored in a subfile onto a proper disk using a subroutine that reformats the data from the input format to a retrieval format. Some data is stored in character

Table 7.4b Formation and Lithology Data (Card Types 08, 35 and 36)

form, and other in hexadecimal form for computational purposes and ease of use in retrieval jobs, and to allow easy access to select specified data from the Database. The stored data in the Database is indexed by a small file of the Database Index on a permanent mounted device. The index file contains the data storage allocation tables as a multirecord header; a work record entry is stored for each distinct catalogue number on the Database. The individual entry records of the catalogue number are created by a routine in the second stage update portion of the maintenance subsystem (**Chart 7.1**).

The retrieval subsystem consists of a translator and a retrieval program. The translator converts a retrieval problem coded in a retrieval language into an adequate format for the retrieval program to read the master files, then selects, manipulates, and outputs data in accordance with the problem. It includes some general purpose manipulation capabilities, such as statistical calculations, counts, sums, ratios, etc.

A dictionary section is provided for the user to describe the information needed for coding retrieval requests, description of sets, nested sets, and items. Master file data items are referred to with standard dictionary names in set and nested set order.

7.2.2.3 The Application Subsystem

The **application subsystem** makes use of the retrieved data to prepare geological and geophysical maps, cross sections, graphs, etc.

The effectiveness of the Petrodata system actually depends on the efficiency of the technical and administrative procedures and regulations that control the overall process.

The stored data on the database retrieval subsystem are processed through the application subsystem, using the Calcomb plotter (or something similar), to perform a series of programs such as **IPI MAP** (isopachs, isobaths, isoporosity maps, etc.); **IPI GRAPH** (graphic analysis); **SONACROSS** (geological cross sections defining lithological characteristics, formation tops, perforation intervals of tubing, etc.); and **SONASEIS** (seismic lines and shot point locations). The computed cross sections can provide a visual display via the plotter to show lithological features in a connected cross section path including up to 20 wells.

The **SONASEIS, SONACROSS, IPIMAP**, and **IPI GRAPH** can be converted to a plot tape before using the Calcomb plotter.

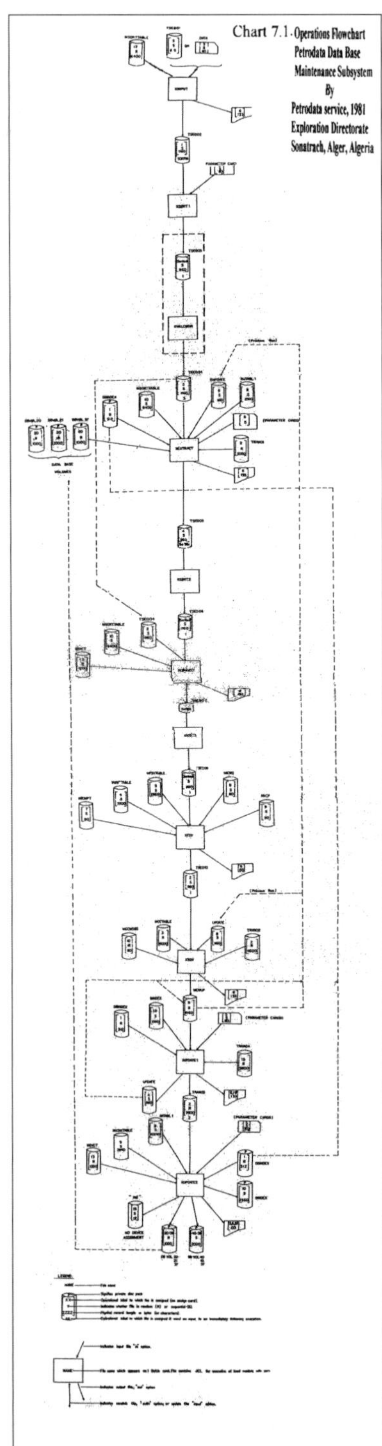

Chart 7.1 Operations Flowchart Petrodata Data Base Maintenance Subsystem By Petrodata service, 1981 Exploration Directorate Sonatrach, Alger, Algeria

7.3 Economic Aspects

Input well data are accumulated in files from service companies over many years, and building up an information bank or database electronically is of extreme importance as it avoids duplication of efforts and saves more than 70% of the professional's time otherwise spent in nonprofessional work. The system should be considerably more effective than the traditional methods in performing certain geological tasks. Specific economic advantages could be gained by applying the Petrodata system, because it serves several professional and technical needs in a much reduced period of time. The Petrodata database retrieval system is quite a sophisticated system that needs sufficient technical personnel to carry out the required jobs in the proper time.

With regard to the efficiency of the Petrodata system, it is advisable to use catalogue numbers for the identification of wells when dealing with any of the three subsystems, instead of depending on the geographic coordinates. Also, it is necessary to select the wells in a certain regional trend, in order for each of the input well data files to be directly processed by the three subsystems successively.

The following questions should be asked before using the Petrodata system:

1. Is there enough budget and are there enough individuals assigned to carry out all the procedures of the three subsystems?
2. How is the system evaluated commercially/ economically? What are the specific advantages?
3. Is there an advantage in the quality of the analysis and in reducing the exploration and production costs that would make the system commercially superior to an apparently cheaper system that gives poorer quality results?
4. Alternatively, is the Petrodata system capable of processing data faster than others, which might enable more rapid evaluation of data, quick and accurate decisions, and consequently more rapid development of resources—all of which would have an obvious commercial advantage?

7.4 Recommendation

Experienced geologists and geophysicists should take the responsibility of analyzing and inputting well data on the data files under the direct supervision of a Petrodata system analyst to guarantee consistency of quality and availability of correct data. Actually, the authorized supervisor keeps in harmony the progress of the validation/correction of well data files and jobs processed at the computer lab.

7.5 A Type Computerized Montage Drilling Report of an Exploratory Well (Boumahni Well "Bi-1")

In 1983, a split computer program documentation, known as the Tape Library Management Program (TALIMAN) was retrieved from the Petrodata system, as an online program in extended Fortran to maintain an up-to-date tape library file. It was applied as a quick method for the preparation of a computed montage report of a proposed exploration well.

The **Boumahni well (Bi-1)** is an exploratory well, drilled and completed by Sonatrach, in association with Sunoco, northeast of the Saharan Platform; it is given as an example for a type computerized montage report; it is located in the Dait El-Kelga permit at longitude 04° 44′ 51″E and latitude 32° 32′ 54″N, about 70 km to the ENE of the Oued Noumer petroleum field (ONR-1), the Ghardia well (GA-1) to the NNW, and the Hassi Boukhellala (HBK-1) to the SWS (see Fig. 7.2).

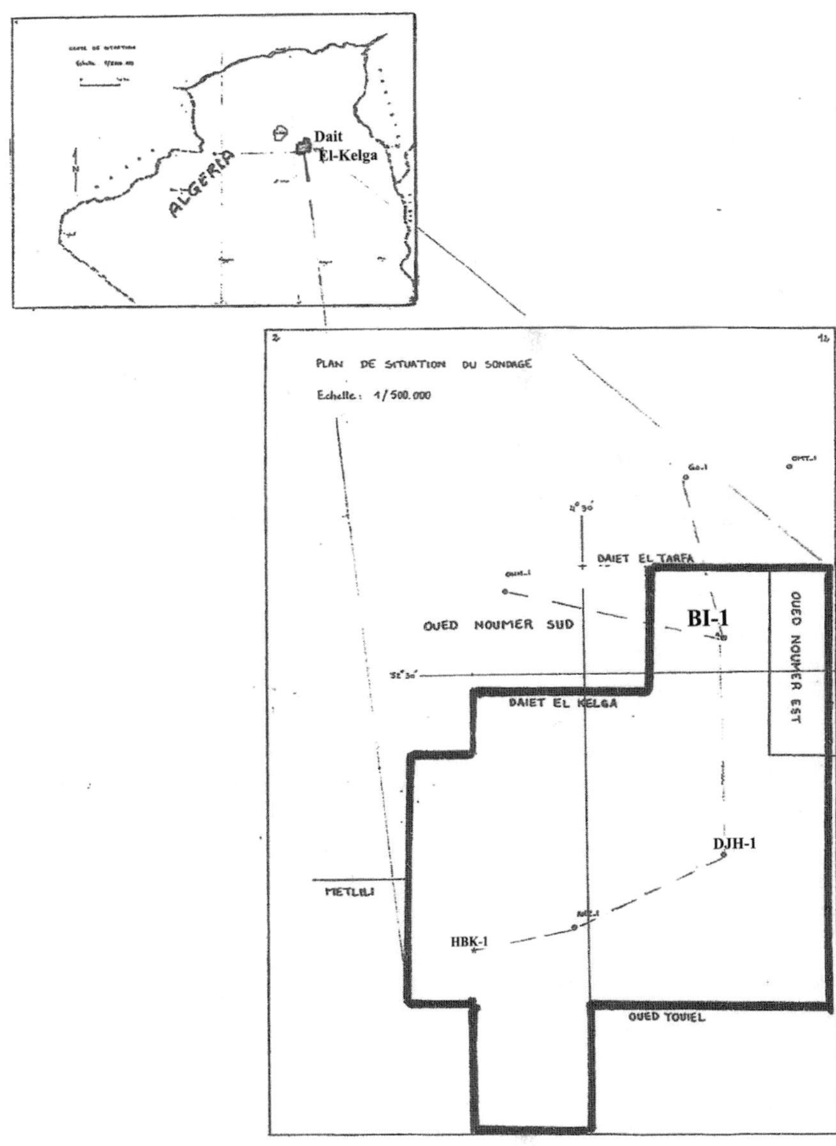

Fig. 7.2 A key map for the proposed Boumahni well (BI-1), NE of the Saharan Platform, Algiers

Fig. 7.3 A structural cross section passing by Ga-1 and HK-1 and the proposed location of Bi-1

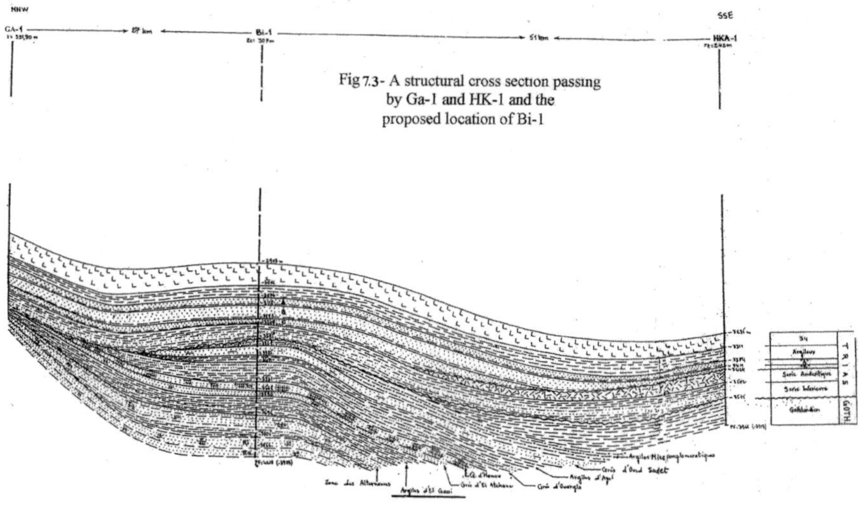

Table 7.5 A comparison table showing different formations encountered at BI-1, and nearby wells

TABLEAU DE COMPARAISON	BOU MAHNI-1 / BI-1 (Zt=307 m) PREVISIONS PROF	CÔTES	EPAIS	GUERRARA-1 / Ga-1 (Zt=332 m) PROF	CÔTES	EPAIS	HASSI BOU KHELLALA / HBK-1 (Zt=326 m) PROF	CÔTES	EPAIS
TERT. — MIO PLIOCENE	Surf.	+307	18	0	+332	146	0	+326	38
TERT. — EOCENE	88	+219	64	—	—	—	—	—	—
CRETACE — SENONIEN — CARBONATE	152	+155	299	149	+183	492	45	+281	212
CRETACE — SENONIEN — ANHYDRITIQUE	451	−144	92				257	+69	82
CRETACE — SENONIEN — SALIFERE	543	−236	90						
CRETACE — TURONIEN	633	−326	48	641	−309	72	339	−13	86
CRETACE — CENOMANIEN	681	−374	182	713	−381	251	425	−99	175
CRETACE — ALBIEN	863	−556	431	964	−632	316	600	−274	505
CRETACE — APTIEN	1294	−987	24	1280	−948	27	1105	−779	25
CRETACE — BARREMIEN	1318	−1011	331	1307	−975	310	1130	−804	345
CRETACE — NEOCOMIEN	1649	−1342	192	1617	−1285	225	1475	−1149	139
JURASSIQUE — MALM	1841	−1534	245	1842	−1510	375	1614	−1288	267
JURASSIQUE — DOGGER — ARGILEUX	2086	−1779	224	2217	−1885	90	1881	−1555	201
JURASSIQUE — DOGGER — LAGUNAIRE							1935	1609	54
JURASSIQUE INFERIEUR — LIAS SUP — ARGILEUX.S	2310	−2003	27	2307	−1975	50	2083	−1757	9
JURASSIQUE INFERIEUR — LIAS SUP — CARBONATE	2337	−2030	37						
JURASSIQUE INFERIEUR — LIAS MOY — ANHYDRITIQUE	2374	−2067	187						
JURASSIQUE INFERIEUR — LIAS MOY — SALIFERE	2561	−2254	45						
JURASSIQUE INFERIEUR — LIAS INF — HORIZON 'B'	2606	−2299	35	2559	−2227	37	2414	−2088	30
JURASSIQUE INFERIEUR — LIAS INF — S1 + S2	2641	−2334	210	2596	−2264	249	2444	−2118	156
JURASSIQUE INFERIEUR — LIAS INF — S3	2851	−2544	256	2845	−2513	170	2600	−2274	300
JURASSIQUE INFERIEUR — ARGILEUX I/O2	3107	−2800	103	3015	−2683	75	2900	−2574	66
TRIAS — S4	3210	−2903	111	3090	−2758	131	2966	−2640	63
TRIAS — ARGILEUX	3321	−3014	62	3223	−2889	31	3029	−2703	27
TRIAS — GRES "A" (T2)	3383	−3076	42	3252	−2920	40	3056	−2730	40
TRIAS — GRES "B" (T1)	3425	−3118	65	3306	−2974	50	3096	−2770	42
TRIAS — GRES "C" (T1)	3490	−3183	25	3356	−3024	44			
TRIAS — SERIE INFERIEURE	3515	−3208	87	3400	−3068	87	3138	−2812	62
GOTHLANDIEN							3200	−2874	71
ORDOVICIEN — DALLE DE M'KRATTA							3271	−2945	25,5
ORDOVICIEN — ARGILES MICROCONGLOM	3602	−3295	32				3296,5	−2970,5	17,5
ORDOVICIEN — GRES D'OUED SARET	3634	−3327	58				3314	−2988	≥1
ORDOVICIEN — ARGILES D'AZEL	3692	−3385	26						
ORDOVICIEN — GRES D'OUARGLA	3718	−3411	110						
ORDOVICIEN — QUARTZITES DE HAMRA	3828	−3521	47						
ORDOVICIEN — GRES D'EL ATCHANE	3875	−3568	22						
ORDOVICIEN — ARGILES D'EL-GASSI	3897	−3590	106	3487	−3155	27			
CAMBRIEN — ZONE DES ALTERNANCES	4003	−3696	109	3514	−3182	≥23,8			
CAMBRIEN — ZONE Ri	4112	−3805	61						
CAMBRIEN — ZONE Ro	4173	−3866	37						
CAMBRIEN — ZONE R2									
CAMBRIEN — ZONE R3									

The main objective is the Triassic shaley sandstone, which is oil productive at the nearby wells of Draa El-Tamra (DRT-1) and the Haniet el-Mokta oil fields. The Boumahni well is located in the Oued M'ya basin where the Triassic reservoir is well developed and is overlain by salt deposits. The seismic isochrone map shows that the well is located on a north/south trend of an anticline with a closed surface area of $42\,km^2$ and vertical closure of a Triassic reservoir (the middle Triassic "B" series) of $14\,m$ (equivalent of $50\,m$). The fractured quartzitic sandstone of the Cambrian and the Upper and Middle Jurassic limestone are secondary objectives. Fig. 7.3 shows a structural cross section among wells GA-1, Bi-1, and HKA-1.

Table 7.5 shows a comparison between the estimated depths of different formations for the proposed well of Bi-1 and the actual depths of the nearby wells of Ga-1 and HBK-1 according to electric log and seismic correlation.

Fig. 7.4 shows the lithology log and casings program for the proposed (Bi-1) well.

A montage report was submitted in a Fortran program, together with a printout of the basic data, e.g., general well data, formation tops, and subsea levels of nearby wells.

7.6 Appendix A

Fortran program and output in Plates 7.1a, b; 7.2a–c.

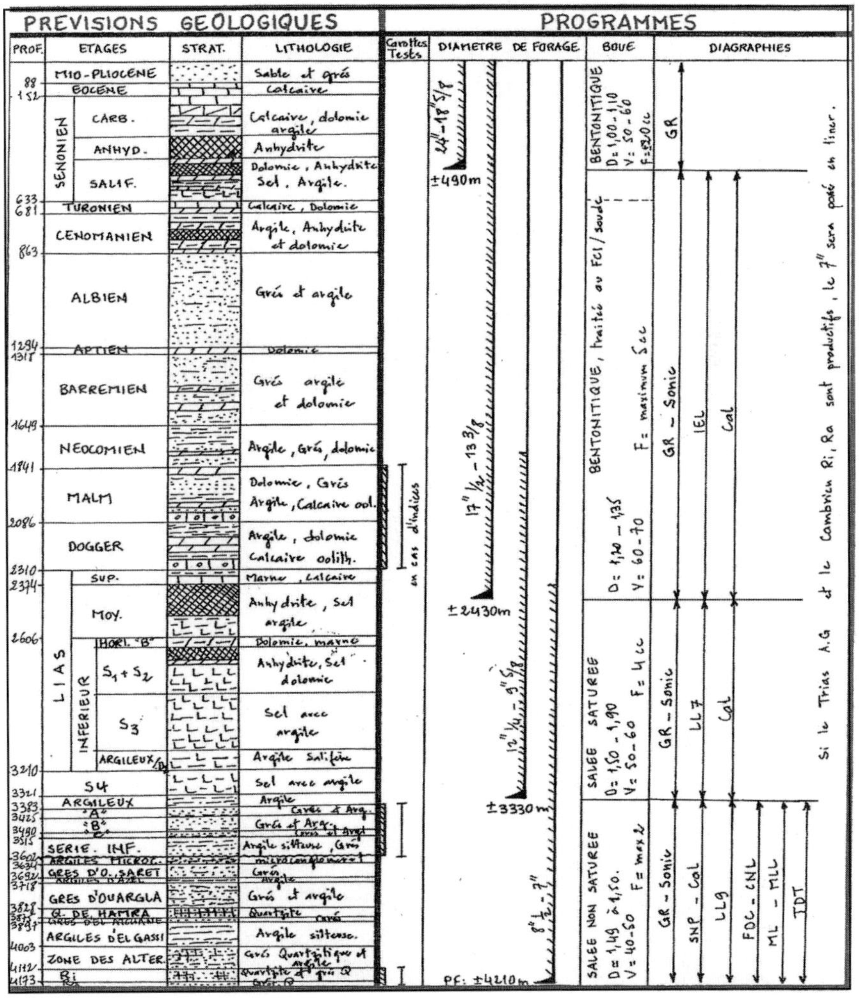

Fig. 7.4 A composite lithologic well log and the design of Casings of the proposed BI well

Plate 7.1a-1

```
!5. APR .00, '63  IO=0195
tEPZM501, BUSSFAKRIDpoi-Aoi-BO1S(13005.DBRS),7,,          .FAKHRY
r (CORE, 55),(TIME, 10),(UO, 1.000)


DE/F505R04 TO LP;K)
   1.000  C              THIS PROGRAM READS GENERAL
   2.000  C              DATA OF AN IMPLANTATION
   3.000  C              EXPLORATORY WELL
   4.000  C
   5.000  C
   6.000  C
   7.000  C
   8.000  C      INTEGER DATE(2),NAME(o),ABREV(6,2),CONCN(4),DIR,GENER(59),
   9.000         1CATL(6),LAT,LONG,MAIN(6),SECOND(9),SITUAT(15),STRUCT(15),
  10.000         2PROF(3),SHOT(2),DIRN,LATTD,LONGIT
  11.000         DIMENSION ZT(6),SIGLE(2)
  12.000         REAL ZS,TDPTH(6),ROTABL,TFORM(6,10),SUBSEA(10),ELEV,T,FDPTH
  13.000  C
  14.000  C
  15.000  C
  16.000         READ(15)(000)DATE,IQATL,NAME,SIGLE,CONCN,LATTD,LONGIT,DIRN,
  17.000         1ZS,ROTABL,MAIN,SECOND,SITUAT,STRUCT,GENER,PROF,SHUT
  18.000    1000 FORMAT(2A4,I6,6A4,2A4,4A4,2IA,A1/2(F6.1),6A4,9A4/2(15A4,/),
  19.000         13(20A4,/),10A4/3A4,2A4)
  20.000         WRITE(108,2000)DATE,ICATL,NAME,SIGLE,CONCN,LATID,LONGIT,DIRN,
  21.000         1ZS,ROTABL,MAIN,SECOND,SITUAT,STRUCT,GENER,PROF,SHOT
  22.000    2000 FORMAT('1',2AX,'1 COMPUTED IMPLANTATION REPORT  FOR'/21X,'AN EXPLOR
  23.000         1ATORY PETROLEUM WELL'/31X,'FAKHRY METHOD-'//35X,'BY'//30X,'FAKHRY
  24.000         2 A-ASSAD'//3IX,'DATE--',2A4//5X,'CATALOGUE NUMBER-',I6,//,X,'WELL'
  25.000         3, NAME-',6A4//5X,'ABREVIATION ',2,4//5X,'CONCESSION-',4A4//5X
  26.000         4,'GEOGRAPHIC LOCATION, LATITUDE ',TD/25X,'LONGITUDE ',I6,A1//5X
  27.000         5,'ZS-',F6-,2NNS,25X,'ZT-',F6.1,2NMS//5X,'MAIN  OBJECT-',6A4
  28.000         6//5X,'SECONDARY OBJECT-',9A4//5X,'SITUATION-',15A4//5X,'STRUC
  29.000         7TURE-', 15A4//5X,'GENERAL GEOLOGY,'//,2(1X,20A4,/),2X,10_4,///
  30.000         02A,'SEISMIC PROFILE-',3A4,25X,'SHOT POINT-',2A4///)
  31.000  C
  32.000  C          THIS PROGRAM READS GENERAL WELLDATA
  33.000  C
  34.000  C               (1) GENERAL WELL DATA
  35.000  C
  36.000      25 WRITE(108,2100)
  37.000    2100 FORMAT('1','1',X,'GENERAL WELL DATA OF NEARBY WELLS',///)
  38.000      50 WRITE(104,2200)
  39.000    2200 FORMAT(2X,'CATALO  ABREV-          WELL NAME         ZS      ZT-  TO
  40.000         1TAL    LATIO  LONG-E-/2X,'NUMBER  TATION--,',
  41.000         23UX,'DEPTH  -VOE  ITUD-N',/41X,'MS',9X,'MS',//)
  42.000         I=I
  43.000      10 READ(16,1100,END=20)CATL(I),(ABREV(I,J),J=1,2),NAME,ZS,ELEVAT,
  44.000         1SDPTH,LAT,LONG,DTR
  45.000    1100 FORMAT(I6,2A4,6A4,F5.1,2(F6.1),2I6,A1)
  46.000      00 WRITE(108,2300)CATL(I),(ABBEV(I,I),J=,2),NAME,ZS,ELEVAT,FDPTH,
  47.000         1LAT,LONG,DTR
  48.000    2300 FORMAT(2X,I6,2A4,6A4,F5.1,2X,2(F6.1,2X),I6,1X,I6,A1///)
  49.000         I=I+1
  50.000         GO TO 10
  51.000      20 CONTINUE
  52.000  C
  53.000  C          (2)FORMATION TOPS
  54.000  C          -DEPTHS OF NEARBY WELLS
  55.000  C
  56.000  C
```

Plate 7.1a-2

```
 7.000      70 WRITE(108,2400)
 8.000    2400 FORMAT('1',10X,'FORMATION TOPS AND SUBSEA DEPTHS OF'//10X,
 9.000       1'NEARBY WELLS'///)
 0.000 C
 1.000 C
 2.000      80 WRITE(108,2500)
 3.000    2500 FORMAT(2X,'CATALOG-ABREV-   ZT(MS) M.OPL- EOCENE CRETAC,APTIAN,
 4.000       1 JURAS- TRIAS- TR-SD- TR-LOW- PALEO- TOTALE '/
 5.000       23X,'NUMBER IATION',11X,'IODENE',10X,'EOUS',11X,'--SIC   ',1X,
 6.000       3'.C-TOP',2X,'STONG',2X,'SERIES',2X,'..ZOIC',2X,'DEPTH'////)
 7.000 C
 8.000 C
 9.000         I=)
 0.000      15 READ(17,120.,END=30)CATL(I),(ABREV(I,J),J=1,2),ZT(I),
 1.000       .(TFORM(I,L),L=1,10)
 2.000    1200 FORMAT(I6,2A4,11F6.1)
 3.000      90 WRITE(108,2600)CATL(I),(ABREV(I,J),J=1,2),ZT(I),
 4.000       1(TFORM(I,L),L=1,10)
 5.000    2600 FORMAT(2X,I6,2A4,2X,11(2X,F6.1),//)
 6.000         I=I+1
 7.000         GO TO 15
 8.000
 9.000 C     (3)SUBSEA DEPTHS OF
 0.000 C         NEARBY WELLS
 1.000 C
 2.000 C
 3.000 C
 4.000      30 CONTINUE
 5.000      95 WRITE(108,2550)
 6.000    2550 FORMAT('1',2X,'CATALOG-ABREV- ZT(NS) MIOPLI EOCENE CRETAC,APTIAN
 7.000       1 JURAS- TRIAS- TR-SD- TR-LOW- PALEO- TOTAL-'/
 8.000       23A,'NUMBER IATION',11X,'IOCENE',10X,'EOUS',11X,'--SIC   ',1X,
 9.000       3'IC=TOP',2X,'STONG',2X,'SERIES',2X,'--ZOIC',2X,'DEPTH'///)
10.000         I=1
 1.000      35 CONTINUE
 2.000         DO 40 J=1,10
 3.000     100 IF (ZT(I).GT.TFORM(I,J)) GO TO 3
14.000         SUBSEA(J)=TFORM(I,J)-ZT(I)
15.000       3 SUBSEA(J)=ZT(I)-TFORM(I,J)
16.000
17.000      40 CONTINUE
18.000         WRITE(108,2700)CATL(I),(ABREV(I,J),J=1,2),ZT(I),(SUBSEA(J),J=1,10)
```

```
72.000
75.000 C  CONT(3)SUBSEA DEPTHS OF
80.000 C         NEARBY WELLS
81.000 C
82.000 C
83.000 C
84.000      30 CONTINUE
85.000      95 WRITE(108,2550)
86.000    2550 FORMAT('1',2X,'CATALOG-ABREV- ZT(NS) MIOPLE EOCENC CRETAC,APTIAN
87.000       1 JURAS- TRIAS- TR-SD- TR-LOW- PALEO- TOTAL-'/
88.000       23X,'NUMBER IATION',11X,'IOCENE',10X,'EOUS',11X,'--SIC   ',1X,
89.000       3'IC=TOP',2X,'STONE',2X,'SERIES',2X,'--ZOIC',2X,'DEPTH'///)
90.000         I=1
91.000      35 CONTINUE
92.000         DO 40 J=1,10
93.000     100 IF (ZT(I).GT.TFORM(I,J)) GO TO 3
94.000         SUBSEA(J)=TFORM(I,J)-ZT(I)
95.000       3 SUBSEA(J)=ZT(I)-TFORM(I,J)
96.000
97.000      40 CONTINUE
98.000         WRITE(108,2700)CATL(I),(ABREV(I,J),J=1,2),ZT(I),(SUBSEA(J),J=1,10)
99.000    2700 FORMAT(2X,I6,2X,2A4,11(2X,F7.1),//)
100.000        I=I+1
101.000        IF (I.GT.6) GO TO 95
102.000        GO TO 35
103.000     45 STOP
104.000        END
```

PROCESSING TERMINATED

Plate 7.1b

A COMPUTED IMPLANTATION REPORT FOR
AN EXPLORTORY PETROLEUM WELL
"FAKHRY METHOD"

BY

FAKHRY A. ASSAD

DATE--JAN03, 83

CATALOGUE NUMBER--471837

WELL NAME--BOUMAHNI-1

ABREVIATION. B1-1

CONCESSION--DAIT ELKELGA

GEOGRAPHIC LOCATION--LATITUDE 323254

LONGITUDE 44151E

25=, 340,6MS   2T=, 307,0MS

MAIN OBJECT--TRIASSIC SHALY S.ST.

SECONDARY OBJECT--UPPER AND MIDDLE JUR.L.ST.

SITUATION--ENE ONNI-BIS FOR 70 KMS.

STRUC TURE--SEISMIC REFLEXTION ON TOP TRIAS

GENERAL GEOLOGY

B1-1 IS SITUATED IN OUED HIYA BASIN--TRIASSIC SHALY S.ST RESERVOIR IS WELL
DEVELOPED--SALT FORMATION AS A CAP ROCK--THE SUBCROP MAP SHOWS THE TRIASSIC
OVERLIES UNCONFORMABLY THE ORDOVICIAN.

SEISMIC PROFILE--72A TI-23-0   SHOT POINT--148-A

Plate 7.2a

A Program of General Well Data
of Nearby Wells

```
1 -   1.000 JAN03, 83471827BOUMAHNI-1          B1-1    DAIT ELKELGA    323254044151E
2 -   2.000   3000  3070TRIASSIC SHALY S.ST.   UPPER AND MIDDLE JUR.L.ST.
3 -   3.000 ENE ONR1-BIS FOR 70 KMS.
4 -   4.000 SEISMIC REFLEXTION ON TOP TRIAS
5 -   5.000 B1-1 IS SITUATED IN OUED HIYA BASIN--TRIASSIC SHALY S.ST RESERVOIR IS WELL
6 -   6.000 DEVELOPED--SALT FORMATION AS A CAP ROCK--THE SUBCROP MAP SHOWS THE TRIASSIC
7 -   7.000 OVERLIES UNCONFORMABLY THE ORDOVICIAN.
8 -   8.000 -72.TI-.7-6 --148-A
```

NAME = GWDATA

```
1 W   1.000 471607 GA-1    GHERRARA-1              3290  3320 35390324037041121E
2 W   2.000 471676 ONN-1   OUED NOUMER NORD-1      3540  3610 3172032284A642136E
3 W   3.000 471895 DJH-1   DJERRAH-1               2620  2690 37920321415044434E
4 W   4.000 471026 ONR-5   OUED NOUMER-5           3950  4010 2834037240404004 8F
5 W   5.000 471725 HKA-1   HANIET EL-NOKTA-1       2350  2420 395403207160459320
6 W   6.000 471624 HBK-1   MASSI BOUKHELLALA-1     3180  3260 331503200230417491E
```

NAME = FORMDATA

```
1 W   1.000 471607 GA-1    3320  0000  1490  2500 12000 18420 25960 32520 34200 34670 35390
2 W   2.000 471676 ONN-1   3610  0000  0000  1220 04640 15720 22440 28940 30610 31210 31720
3 W   3.000 471895 DJH-1   2090  0000  0000  1635 08850 13550 20450 26300 7700  28250 37920
4 W   4.000 471026 ONR-5   4010  0000  0000  1930 .7960 18470 27700 34000 35980 36690 28340
5 W   5.000 471725 HKA-1   2420  0000  0000  1160 1406  26380 28830 36260 37440 39170 39540
6 W   6.000 471624 HBK-1   3260  0000  0000  1510 11050 10170 24430 30560 31590 32000 33150
```

Plate 7.2b

GENERAL WELL DATA OF NEARBY WELLS

CATALOG NUMBER	ABREVIATION	WELL NAME	ZS MS	ZT MS	TOTAL DEPTH	LATITUDE-N	LONGITUDE-E
471007	GA-1	GUERRARA-1	329.0	332.0	3539.0	324637	44121E
471678	ONN-1	OUED NOUHER NORD-1	354.0	361.0	3172.0	322844	42136E
471893	DJN-1	DJERH-N-1	262.0	269.0	3792.0	321415	44434E
471628	ONR-5	OUED NOUHER-5	395.0	401.0	2864.0	322404	40048E
471723	HKA-1	HANIET EL MOKTA-1	235.0	242.0	3964.0	320716	45932E
471824	HBK-1	HASSI BOUKHELLALA-1	318.0	326.0	3315.0	320623	41749E

Plate 7.2c

CATALOG NUMBER	ABREVIATION	ZT(MS)	MIDPA-LEOCENE	EOCENE	CRETAC-EOUS	APTIAN	JURAS-SIC	TRIAS-IC-TOP	TR.SD-STONE	TR.LON.SERIES	PALEO-ZOIC	TOTAL-DEPTH
471007	GA-1	332.0	0.0	149.0	250.0	1280.0	1842.0	2596.0	3252.0	3420.0	3487.0	3539.0
471678	ONN-1	361.0	0.0	0.0	82.0	944.0	1572.0	2244.0	2891.0	3061.0	3121.0	3172.0
471893	DJN-1	269.0	0.0	0.0	163.5	825.0	1355.0	2045.0	2630.0	2700.0	2825.0	3792.0
471628	ONR-5	401.0	0.0	0.0	193.0	1396.0	1947.0	2770.0	3460.0	3598.0	3669.0	3834.0
471723	HKA-1	242.0	0.0	0.0	116.0	1506.0	2038.0	2888.0	3626.0	3744.0	3817.0	3954.0
471824	HBK-1	326.0	0.0	0.0	150.0	1105.0	1617.0	2443.0	3056.0	3139.0	3200.0	3315.0

SUBSEA DEPTHS OF

NEARBY WELLS

CATALOG NUMBER	ABREVIATION	ZT(MS)	MD-PLIO-EOC-EOCENE	EOCENE	CRETAC-EOUS	APTIAN	JURAS-SIC	TRIAS-IC-TOP	TR.SD-STONE	TR.LOW-SERIES	PALEO-ZOIC	TOTAL-DEPTH
471007	GA-1	332.0	+332.0	+183.0	+82.0	-948.0	-1510.0	-2264.0	-2920.0	-3088.0	-3155.0	-3207.0
471678	ONN-1	361.0	+361.0	+361.0	+339.0	-583.0	-1211.0	-1883.0	-2530.0	-2700.0	-2760.0	-3811.0
471093	DJF-1	269.0	+269.0	+269.0	+105.5	-556.0	-1086.0	-1776.0	-2361.0	-2431.0	-2556.0	-3523.0
471628	ONR-5	401.0	+401.0	+401.0	+288.0	-995.0	-1546.0	-2369.0	-3059.0	-3197.0	-3268.0	-3433.0
471723	HKA-1	242.0	+242.0	+242.0	+120.0	-1264.0	-1796.0	-2646.0	-3384.0	-3508.0	-3575.0	-3709.0
471824	HBK-1	326.0	+326.0	+326.0	-176.0	-779.0	-1291.0	-2117.0	-2730.0	-2813.0	-2876.0	-2989.0

STOP 0

References

Assaad FA (1978) Introduction to Petrodata system III. Seminaire Nationale des Sciences de la Terre, 27–29 Nov., "Section de Geologie Petrolier," INH, Boumerdas, Algeria; also by Sonatrach at the OAPEC conference in Kuwait, 1979.

Assaad FA (1978) Preliminary geological application of Petrodata system for Petroleum Exploration Directorate of Algeria. Sonatrach (Internal Report), Algiers, Algeria.

Assaad FA (1978) A computerized provisional listing of catalogue numbers for exploration and production wells implanted in Algeria during the period of Mar 1972 to Dec 1977. Sonatrach (Internal Report), Algiers, Algeria.

Assaad FA (1976) Essai d'homogeneisation des correlation lithostratigraphiques des formations Paleozoiques en Algerie. Sonatrach (Internal Report), Algiers, Algeria.

Glossary

Plate Tectonic Features—Definitions (see Chap. 2)

- **Pangaea**—A supercontinent that existed from about 200 to 300 million years ago and included most of the continental crust of the earth from which the present continents were derived by fragmentation and continental displacement; Pangaea is the hypothetical single landmasss that is believed to have split into two large fragments: Laurasia in the north and Gondwana in the south.
- **Rift Valley**—An elongated depression, trough, or graben in the earth's crust, bounded on both sides by normal faults or fault zones of approximately parallel strike and occurring on the continents or under the oceans; e.g., the Red Sea and the African Rift Valley are commonly the sites of volcanoes and the locus of much earthquake activity.
- **Subduction Zone**—a large scale narrow region in the earth's crust where, according to plate tectonics, masses of the spreading oceanic lithosphere bend downward into the earth along the leading edges of converging lithospheric plates, where they slowly melt at about 400 mi (600 km) deep and become re-absorbed.
- **Plate**—A segment of the lithosphere which has little volcanic or seismic activity but is bounded by almost continuous belts (or plate margins) of earthquakes and, in most cases, by volcanic activity and young subsea or subaerial mountain chains. Most earth scientists think there are currently seven large, major plates: the African, Antarctic, Eurasian, Indo-Australian or Indian, North American, Pacific, and South American plates. There are also several smaller plates, e.g., the Arabian, Caribbean, Cocos, Nazca, and Philippine plates. The positions of the boundaries of some present-day plates are disputed, particularly within and adjacent to collision zones, e.g., the Alpine-Himalayan belt.
- **Plate Motions**—The movement of tectonic plates is expressed in terms of rotations relative to an Euler pole. Such motions during the last 200 million years have been determined mainly from magnetic anomaly patterns in the ocean basins. Determination of older motions is based on palaeomagnetic studies and evidence for continental collisions and separations.
- **Plate Tectonics**—The unifying concept that has drawn continental drift, sea-floor spreading, seismic activity, crustal structures, and volcanic activity into a coherent model of how the outer part of the earth evolves. The theory proposes a model of the earth's upper layers in which the colder, brittle, surface rocks form a shell (the lithosphere) overlying a much less rigid asthenosphere. The shell comprises several discrete, rigid units (tectonic plates), each of which has a separate motion relative to the other plates. The plate margins are most readily defined by present-day seismicity, which is a consequence of the differential motions of the individual plates. The model is a combination of continental drift and sea-floor spreading. New lithosphere plates are constantly forming and separating, and so being enlarged, at constructive margins (ridges); while the global circumference is conserved by the subduction and recycling of material into the mantle at destructive margins (trenches). This recycling results in andesitic volcanism and the creation of new continental crust, which has a lower density

than oceanic crust and is more difficult to subduct. Many features of the Earth's history are explicable within this model, which has served as a unifying hypothesis for most of the earth sciences. Previous mountain systems are now recognized as the sites of earlier subduction, often ending with continental crustal collisions; the movement of plates has been used with varying success in interpreting orogenic belts as far back as the early Proterozoic. Plate motions are driven by mantle convection and are likely to have occurred throughout earth history, although the resultant surface features are likely to have changed with time.

Plate Margin (Plate Boundary)—The boundary of one of the plates that form the upper layer (the lithosphere) and that together cover the surface of the earth. Plate margins are characterized by a combination of tectonic and topographic features, e.g., oceanic ridges, young fold mountains, and transform faults. Plate margins are of three main types: (a) constructive margins, where newly created lithosphere is being added to plates which are moving apart at oceanic ridges; (b) convergent margins, which can be either destructive margins, where one plate is carried down into the mantle beneath the bordering plate at a subduction zone, or a collision zone, where two island arcs or continents or an arc and a continent are colliding; or (c) conservative margins, where two plates are moving in opposite directions to each other along a transform fault. All three margins are seismically active, with volcanic activity at constructive and destructive margins. Some plate margins exhibit features of more than one of the three main types and are known as combined plate margins.

Examples of Plate Junctures (see Chap. 2)

– **Transform Plate Junctures**—The San Andreas transform fault is an enormous seismic energetic structure, 800 km in length, with about 240 km of relative horizontal movement that causes earthquakes in California. Such transform faults are parallel to each other, extend thousands of kilometers, and are perpendicular to the ridge. They are most numerous in mid-ocean ridges, e.g., the Mid-Atlantic Ridge.
– **Divergent Junctures**—Rift structures represent deep faults between two divergent structures where

basaltic magma is pouring out along normal faults, forming volcanic domes or plate-like structures which extend hundreds and thousands of kilometers with considerable amplitude of vertical movement (2–4 km), e.g., the large Afro-Arabian rift belt, about 6,500 km long.

– **Juncture**—A point or line of joining or connection
– **Junction**—A point where two or more passageways intersect horizontally or vertically.
– **Geosynclinal Prism**—a load of sediments which accumulates in the down-warped area of a geosyncline.
– **Subduction**—The sinking of one crustal plate under another as they collide

General Stratigraphic and Structural Terms

- **Flysch**—Widespread deposit of sandstones, marls, shales, and clays which lie on the north or south portions of the Alps.
- **Spilite**—A basaltic rock with albitic or sodic feldspars. The albitic feldspar is usually accompanied by autometamorphic minerals, or minerals characteristic of low-grade greenstones such as chlorite, calcite, epidote, etc. "**NB**: spilitization is the transformation process of basalt to spilites by low grade metamorphism."
- **Orogenic Movement**—Mountain making—denoting the process of formation of mountain ranges by folds, faults, upthrusts, and overthrusts, affecting comparatively narrow belts and lifting them into great ridges.
- **Epeirogeny and Epeirogenetic**—Designating the broad movements of uplift and subsidence which affect the whole or large parts of continents and the oceanic basins
- **Epeirogenic Movements—Movements** of the earth's crust which produce and maintain the continental plateaus and the broad depressions which are covered by the sea.
- **Fluvial Deposits**—Produced by a river.
- **Alluvial Deposits**—Made up of alluvium.
- **Alluvial Fan**—A gradually sloping mass of alluvium that widens out like a fan from the place where a stream slows down little by little as it enters a plain.

- **Delta**—An alluvial deposit, usually triangular, at the mouth of a river.
- **Deltaic Deposits**—Sedimentary deposits laid down in a river delta.
- **Nappe**—A large body of rock that has moved forward more than one mil from its original position, either by overthrusting or by recumbent folding.
- **Nappes**—Faulted overthrusted folds.
- **Lacustrine**—having to do with a lake or lakes; found or formed in lakes
- **Terrace**—A geologic formation; any of a series of flat platforms of earth with sloping sides, rising one above the other, as a hillside; or a raised, flat mound of earth with sloping sides.
- **Diapir**—A dome formation in which the rigid top layers have been split open by pressure from an underlying plastic core.
- **Geosyncline**—A very large, trough-like depression in the earth's crust containing masses of sedimentary and volcanic rocks.
- **Geosynclinal Prism**—The load of sediments which accumulates in the down-warped area of a geosyncline.
- **Geosyncline Trough**—A region subsiding over a long time, containing sedimentary and volcanic rocks of great thicknesses, primarily interpreted as an evidence of subsidence.
- **Orogeny**—The formation of mountains through structural disturbance of the earth's crust, especially by folding and faulting.

Foreland—The relatively stable area, lying in shallow water, represented by the continental platform; or the resistant block towards which the geosynclinal sediments move when compressed; in structure, it is the region in front of a series of overthrust sheets.

Sedimentary Basins—A wide, depressed area in which the rock layers all incline toward a center.

- **Appalachian Mountains**—A mountain system in eastern North America; "known for its tea plant leaves used in pioneer times; Appalachian tea plants."
- **Island Arcs**—A group of islands having a curving arc-like pattern. Most the island arcs lie near the continental masses, but inasmuch as they rise from the deep ocean floors, they are not a part of the continents proper.
- **Rif**—Mountain range along the northeastern coast of Morocco, extending from the Strait of Gibraltar to the Algerian border; highest peak 8,000 ft (2,440 m).
- **Rift**—A large fault along which movement is mainly lateral.
- **Mesogeosyncline**—A geosyncline between two continents and receiving clastics from both of them.
- **Deltaic Deposit**—A sediment deposit laid down in a well developed delta; can be characterized by cross bedding, mixture of sand and clay, and remains of brackish matter.
- **Great Rift Valley**—Depression of southwestern Asia and East Africa, extending from the Jordan River across Ethiopia and Somalia to the lakes region of East Africa.
- **Ridge**—A relatively narrow elevation which is prominent because of the steep angle at which it rises.
- **Ophiolite**—A basic igneous rock belonging to an early phase of the development of a geosyncline, and subsequently altered into rocks rich in serpentine, chlorite, epidote, and albite.

Index